THE CODIFICATION OF MEDICAL MORALITY

Philosophy and Medicine

VOLUME 45

The titles published in this series are listed at the end of this volume.

THE CODIFICATION OF MEDICAL MORALITY

Historical and Philosophical Studies of the Formalization of Western Medical Morality in the Eighteenth and Nineteenth Centuries

Volume One: Medical Ethics and Etiquette in the Eighteenth Century

Edited by

ROBERT BAKER

Department of Philosophy, Union College, Shenectady, New York

DOROTHY PORTER

Birkbeck College, University of London

and

ROY PORTER

Wellcome Institute for the History of Medicine, London

KLUWER ACADEMIC PUBLISHERS

DORDRECHT / BOSTON / LONDON

Library of Congress Cataloging-in-Publication Data

The Codification of medical morality : historical and philosophical studies of the formalization of Western medical morality in the eighteenth and nineteenth centuries / edited by Robert Baker, Dorothy Porter, and Roy Porter.

p. cm.

Includes bibliographical references and index.

Contents: v. 1. Medical ethics and etiquette in the eighteenth century.

1. Medical ethics--History--18th century. 2. Medical ethics--History--19th century. I. Baker, Robert, 1937- . II. Porter, Dorothy, 1953- . III. Porter, Roy, 1946- .

[DNLM: 1. Gregory, John, 1724-1773. 2. Percival, Thomas, 1740-1804. 3. Ethics, Medical--history--congresses. 4. History of Medicine, 18th Cent.--congresses. 5. History of Medicine, 19th Cent.--congresses. 6. Philosophy, Medical--history--congresses. W3 PH609 v.45 / WZ 56 C669]

R724.C545 1993

174'.2'09409033--dc20

DNLM/DLC

for Library of Congress 92-49845

ISBN 978-90-481-4193-7

Published by Kluwer Academic Publishers,
P.O. Box 17, 3300 AA Dordrecht, The Netherlands.

Kluwer Academic Publishers incorporates
the publishing programmes of
D. Reidel, Martinus Nijhoff, Dr W. Junk and MTP Press.

Sold and distributed in the U.S.A. and Canada
by Kluwer Academic Publishers,
101 Philip Drive, Norwell, MA 02061, U.S.A.

In all other countries, sold and distributed
by Kluwer Academic Publishers Group,
P.O. Box 322, 3300 AH Dordrecht, The Netherlands.

Printed on acid-free paper

Printed in the Netherlands

TABLE OF CONTENTS

PREFACE

The editors have incurred many debts in preparing this book, and both etiquette and ethics would be contravened if they were not discharged here. Above all, we wish to thank the contributors for so cheerfully complying with our suggestions for preparing their papers for publication and efficiently meeting our schedules. It is thanks to their cooperation that this volume has appeared speedily and painlessly; their revisions have helped to give it internal coherence.

This volume has emerged from papers delivered at a conference on the History of Medical Ethics, held at the Wellcome Institute for the History of Medicine, London, 1 December, 1989. We are most grateful to the Wellcome Trust for having underwritten the costs of the conference, and to Frieda Houser and Stephen Emberton whose organizational skills contributed so much to making it a smoothly-run and enjoyable day.

In addition to the papers delivered at the conference, we are delighted to have secured further contributions from David Harley and Johanna Geyer-Kordesch. Our thanks to them for their eager help.

From start to finish, we have received splendid encouragement from all those connected with the Philosophy and Medicine series, especially Professor Stuart Spicker, and Martin Scrivener at Kluwer Academic Publishers. Their enthusiasm has lightened our load, and expedited the editorial process.

In this age of electronic texts, we owe a special debt of gratitude to Marianne Snowden, and the computing staff at Union College for the months they spent unscrambling and debugging diskettes to form a single, harmonious electronic text. We would also like to thank two Union College students, Elizabeth Lambert and Gregory Palmer, for their assistance in preparing the index.

The present volume is organized into three sections. The first explores the standards and breaches of medical propriety prior to the formalization of medical ethics that can be found in Gregory's lectures and Percival's

R. Baker, Dorothy Porter and Roy Porter (eds), The Codification of Medical Morality, vii–viii.

Code. The second section acquaints the reader with the philosophical frameworks with which Gregory and Percival would have been familiar, given their training in Leiden and Edinburgh. The final section focuses on the lives and writings of Gregory and Percival.

A second volume will take up the development of these and other themes in the nineteenth century, and appropriate both Gregory's idealized sympathetic physician and Percival's conception of a medical code and harness them to the ideal of a self-governing medical profession.

ROBERT BAKER
DOROTHY PORTER
ROY PORTER

INTRODUCTION

Attention to medical ethics has become an integral and essential aspect of modern medicine. No longer can physicians presume that the only judgments facing them are clinical: in many departments of practice, ethical dilemmas are multiplying as never before, and with terrifyingly complex ramifications: personal, legal, professional, administrative, political and from a practical viewpoint, not least financial. Nowadays it is no longer possible to initiate clinical trials and tests or to make therapeutic innovations without extensive attention to their ethical implications. On both sides of the Atlantic, though particularly in North America, such deliberations are often undertaken with considerable bureaucratic formality, and in an increasingly juridical, even legalistic, atmosphere [22]; [73]; [74]; [75]; [76]; [79].

Such developments over the last twenty years have created a certain ambiguity towards earlier attempts to codify proper medical practice and more generally towards the subject of the history of medical ethics. On the one hand, there is the belief that the "founding fathers" of medical ethics guide our paths, or at least *should* do so. Even today, discussion of such principles as the 'sanctity of life' routinely invoke the formulations of Hippocrates [8]; [41]; [85]; [70]; and there remains an inclination to clinch contested questions of professional etiquette by a choice apothegm from Oliver Wendell Holmes or Sir William Osler. In an age of increasing ethical uncertainty, it is sometimes tempting and always reassuring to be able to invoke authorities of old who established the groundwork, and whose wisdom may serve as an ultimate court of appeal [2].

Yet at the same time, the sheer intensity and sophistication of modern developments may also create the temptation to disparage classic early formulations of medical ethics – Hippocrates, Gregory, Percival, the codes of the nineteenth-century medical associations – or at least to dismiss them with a mixture of nostalgia and envy. Those were the days in which it was simple to be wise! The pressures upon the physician were fewer, and (so the tale runs) he won public respect if he showed himself a

R. Baker, Dorothy Porter and Roy Porter (eds), The Codification of Medical Morality, 1–14.

good fellow, a scholar, and a gentleman, as well as a dignified diagnostician or a sturdy sawbones. No one was concerned with patients' rights, or confidentiality, or informed consent; the doctor could get on with practicing medicine [40]; [57].

Both these attitudes – the past erected into an authority, the past as a golden age of nostalgia – certainly sustain interest in the history of medical ethics. But neither of them may make for historical accuracy or understanding. There is, for one thing, a danger of creating a fictive "Hippocrates our contemporary," a dispenser of timeless epigrammatic wisdom – his words, his views, wrenched out of the specific medical, philosophical, and semantic contexts of Greek medicine and society. And there is the no less serious risk of conjuring up a mythic picture, in which what constituted good medical ethics (before our age of complexity and confusion) were neither philosophical principles nor the arid technicalities of pettifogging attorneys, but the good will and wisdom of the character of the dependable practitioner.

Medical issues, however, have never been timeless. Nor was there in reality some golden age of innocence, before what might be condemned as the monstrous modern intrusion of government, lawyers, academics, philosophers, and even 'ethicists'. Close study of the successive formulations of medical ethics demonstrates that it is nothing new for physicians to be confronted with novel and agonizing problems of unexplored biotechnical possibilities and uncertain public response.

Examined with care, the formulations of medical ethics over previous centuries, both theoretical and practical, are revealed to have been as complex and as entangled in philosophical principle, as we feel today's situation to be. For this reason, precise investigation is needed of the texts of successive codifications, the contexts in which they were formulated, and the chain of interplay and influence attending them down the centuries. The pronouncements of medical ethics mean little – or at least their meaning is sure to be misunderstood, the names of their authors taken in vain – unless they are approached and interrogated historically. This volume is offered as a contribution to this task.

It is, in fact, remarkable how little first-hand scholarship has been devoted to researching the history of medical ethics. It is a topic that has been generally neglected by both historians of moral philosophy and historians of medicine. The grand overviews are rather few, being derivative, and in many respects suspect [15]; [25]; [43]; [80]; [81]; [23]; [32]; [65]; [17]; [1]; [71]; [19]; [33]; [69]. A limited amount of fine research has been given over to individual figures and problems, such as the provenance of the Hippocratic Oath, which turns out to have scant connection with any historical Hippocrates [44]; [52]; [45]; [39], and the emergence of professional codes in nineteenth- and twentieth-century North America [10]; [11]; [51]. But in most of this work, the level of scholarship and interpretative sophistication has not been notably high.

One valuable critical input, over the last generation, has come from sociologists of medicine such as Jeffrey Berlant in the United States and Ivan Waddington in Britain. They have argued that formulations such as Thomas Percival's *Medical Ethics* have too often been read out of context and in a literal-minded way, as if they were better or worse contributions to the construction of a perfect code of medical ethics [82]; [13]; [6]; [24]; [34]; [32]; [33]. Traditional "idealist" readings, this critique alleged, accepted ethical codes at face value. Instead, their high-minded intentions of protecting the patient, and their picture of the disinterested physician needed critical interpretation. Such codes, Berlant and Waddington argued, had commonly served as ideological weapons in strategies of professional enhancement, rationalizing a paternalist and monopolistic control over the sick in the name of expertise and benevolence [6]; [82].

At the time of publication, this critical sociological account struck many as challenging and debunking. Iconoclasm readily turns orthodoxy, and by now many of Berlant's and Waddington's main contentions have become absorbed within routine orthodox readings of the history of the medical profession; they are themselves being challenged by historians as lacking in nuance. Perhaps one set of pat answers has been replaced by another. Certainly, the time is ripe for scholars further to explore the problems earlier exposed and to refine and modify the revisionism of the historical sociologists. It is today acknowledged that the eighteenth century constituted a crucial epoch in the crystallization of medical ethics. There is some agreement as to the kinds of factors underlying these developments: rising demand for medicine, the emergence of a more literate, more demanding public in the age of Enlightenment, the advent of a better trained medical profession, many of whom had undergone a philosophically-oriented university education; the growth of new medical institutions, and so forth. But as yet, hardly any attempts have been made to construct a broad survey of the role played by medical ethics in eighteenth-century medicine, philosophy, and society; few studies have appeared examining the key texts in depth; and still fewer efforts have been made to trace medical ethics as a *real presence* down the generations. This volume poses such questions, and ventures certain provisional answers.

It may be asked of seventeenth-century Britain: can we even speak of the presence of medical ethics? That phrase itself was certainly not yet in common usage: one looks up 'medical ethics' in early eighteenth-century medical dictionaries and handbooks in vain [64]; [35]. But if the phrase was lacking, the reality was surely present. Andrew Wear and Harold Cook in particular have recently shown that many sorts of edifying and instructional literature, produced for physicians in pre-1700 England, standardly addressed questions central to any notion of medical ethics: Was the doctor to be motivated by love of money, of fame, of the patient, or of his art? What were the distinguishing marks dividing the reputable doctor from the quack? – and so forth [16]; [3]; [4]; [84].

The same appears true for the eighteenth century. One may argue that Georgian medical practitioners became sucked into the whirlpool of an ebullient market economy [59]; [34]; [48]. One may also listen to the cynicism directed by satirists against the tricks of the profession [63]; [77]. Nevertheless, there is little sign that anyone involved in the philosophy or the practice of medicine in the eighteenth century wished to abandon the claim that, in principle, skilful physicians could be differentiated from incompetent, honest from fraudulent, and that the criteria governing such discriminations were ethical.

It is possible to ask whether such a beast as medical ethics existed in seventeenth-century England. Such a question would be foolish for much of the Continent at that time, since formal discussion of the rectitude and propriety of the conduct of physicians was already deeply entrenched in many legal and academic environments in Italy, France, and the German principalities. In almost every respect, the business of adumbrating and implementing medical ethics, sorting good practice from bad, was more formal, better institutionalized in states of an absolutist tendency, and in jurisprudential traditions where Roman law was powerful. As Johanna Geyer-Kordesch demonstrates, in Prussia, and, more generally, in the German-speaking world, medicine had already established a prominent position as a university discipline [26]; [27]; [28]; [29]. Medical jurisprudence was a prescribed and prestigious academic discipline, shaped by the protocols of administrative law. Medical professors, and medically-expert lawyers, were commonly called upon to resolve judicial issues and to serve the state. Under such conditions, ticklish medical decisions were commonly adjudicated according to conventions of moral and natural law [18].

A somewhat similar sense of the regulation of medicine from above by higher authorities perhaps also obtained in *ancien régime* France. At the regional level, entry into the profession was strictly policed by guild control. Nationally, the Crown gave sanction to the efforts of public bodies such as the Société Royale de Médecine to eliminate improper and exploitative practice. One must not, of course, naively assume that reality operated this way: account must be taken of medical politicking and intra-professional jealousies. Nevertheless, it is clear that the voice of authority lent its weight, in France as in the German principalities, to certain sorts of medical practice as being more respectable, more proper and, ultimately, more legal than others. State authority guaranteed medical propriety and, in this sense, the ethical practice of medicine [66]; [67]; [68].

How far any of this scenario applied to England – indeed, to Britain – is far more disputable. Recent scholarship has been arguing that medicine, in effect, became largely deregulated in Georgian England. Medical corporations such as the Royal College of Physicians grew inactive; neither the Crown, nor Parliament, nor urban corporations, nor local magistrates, exercised themselves very energetically to enforce the rectitude of particular sorts of medical practice [59]. Moreover, under English common law, the courts seem rarely to have been inveigled into adjudicating matters of medical malpractice and malfeasance. This was, it might be suggested, because, especially thanks to the reforms of Lord Mansfield, the legal system increasingly adapted itself to accept the rules of the market, with their fundamental assumption of *caveat emptor*, and became unwilling to interfere with the freedom of trade [7]; [53]; [72]; [73]; [16]; [18].

At the same time, with the Crown awarding patents to the vendors of proprietary nostrums and taking tax revenue from patent medicines, the distinctions between regulars and irregulars grew more murky. This situation was exacerbated, as David Harley's paper emphasizes, by the fact that regulars were more than ready to rush into print, publicly mudslinging against each other [60]; [61]. Overall, it was widely argued that in the flourishing and lucrative "sick trade" ([5], p. 100), regulars had stolen the clothes off the back of the quacks, increasingly giving themselves over to gross and shameless pursuit of commercial profit.

Roy Porter's essay explores the accusations to this effect, levelled against his fellow practitioners by the late eighteenth-century radical

Bristol physician, Thomas Beddoes. Beddoes was certainly not alone in offering what we might call a sociology of what he condemned as the corruption of true, principled medical practice by sordid greed for gain. Doctors, he asserted, had been seduced by the siren strains of market values. To counter this, Beddoes called for greater rectitude amongst the profession, simultaneously demanding higher respect for physicians from the community. Medicine could become ethical only if its social relations were transformed, only if the profession to become more high-minded.

Similar views of the physician's surpassing *duty* were being expressed at the same time by religious moralists, especially evangelicals such as Thomas Gisborne [30]; [62]. Indeed many commentators ventured "sociological" accounts as to why it was not merely unethical, but actually imprudent – crudely, in the long run, bad for business – for doctors to allow themselves to be identified in the public mind too closely with tradesmen. The status, dignity, and authority necessary for a distinguished profession would accrue not from following the laws of the market ever more assiduously, but from adopting the mien of a gentleman.

Indeed, prior to the publication of Thomas Gregrory's *Lectures on The Duties and Qualifications of a Physician* in 1772, in what one might loosely term the "prehistory" of modern medical ethics, at least in the English-speaking world, recommendations as to the proper behavior of doctors were often hard to distinguish from the much broader genre of *advice to gentlemen* purveyed in general conduct manuals. Hence the key question posed by Mary Fissell's essay: What precisely were the historical affinities between the ethics traditionally prescribed for a gentleman, and the emergence of a corpus of specifically medical ethics? Fissell suggests that there are many ways in which medicine had no unique professional ethical code of its own – indeed *needed* none – until the perception of a crisis in long-standing gentlemanly ethics. This came about in the latter part of the eighteenth century, partly with the Evangelical critique of the noble code of honor (and its commitment to unethical abominations such as duelling), and partly because of the ridicule excited by Lord Chesterfield's exaggeratedly cynical reading of the code of a gentleman. In any case, in the era of the Industrial Revolution a more independent and insistent bourgeois identity was itself emerging [56]; [14].

John Pickstone's and Robert Baker's discussions of Thomas Percival both shed light on this question. For Percival himself constitutes a crucial transitional figure. Himself Leiden and Edinburgh educated (in themselves, controversial qualifications – see Porter's examination of Beddoes's strictures against the Edinburgh "medical manufactory") and practicing as a physician in industrializing Lancashire, Percival was in many respects impeccably bourgeois. Hence it comes as no surprise that (as Pickstone contends) some of Percival's broader writings were indeed, as Fissell's argument would predict, criticisms of aristocratic excess, debauchery, and such cruelties as blood sports. In his educational tract for young people, *A Father's Instructions Adapted to Different periods of Life* (1775), Percival was wholeheartedly committed to the inculcation of genteel values and manners [47]; [54]; [50].

The precise amalgam of bourgeois sense and gentlemanly sensibility – perhaps ultimately revealing the desire to be more gentlemanly than those to the manner born – is evident throughout Percival's *Medical Ethics*, as is demonstrated by Robert Baker's helpful textual and contextual analysis. As a member of society, the physician must respect the freedoms of the genteel patient – not least, so as to display his own breeding and liberality. As a physician, however, the practitioner must follow the rules of his own vocation, with its special expertise. Yet he must also, as a man of business, ensure his own financial security and success. Perhaps only a Mancunian would actually be so candid as to have said in so many words that "the profession of physician cannot be supported except as a lucrative one" [55].

As Baker emphasizes, Percival addressed himself to the creation of cordial relations between the different grades of the medical profession, to forestall intra-professional jealousies. Medicine could not be allowed to be sullied by the sordid rivalries of the counter and the market (an ugly situation which had arisen in Percival's Manchester when the medical staff at the Infirmary had been extended in the early 1790s, thus providing the spur to Percival's work) [58].

The paper by Pickstone spells out much of the context for Percival's formulations; it stresses the circumstances which provoked him to write – the frayed Manchester medical politics – and to explain some of Percival's aspirations. But local rivalries and the problems of relating trade and gentility account only partly for Percival's treatise, and for its great success and long-term impact. Another dimension must now be addressed: the formal intellectual tradition within which Percival was operating, that is, the tradition of moral philosophy, most clearly associated with the Scottish universities. These themes are developed by

Larry McCullough, who demonstrates that the major issues raised in the first major work of medical ethics in English, John Gregory's *Lectures*, arose directly from the leading concerns of Scottish moral thinking.

To this point, this Introduction has tacitly been implying a divide between the rather formal, jurisprudential, and philosophical traditions of medical ethical thinking characteristic of the Continent, and the rather pragmatic and socially-oriented tradition typical of England. There is a degree of truth in this dichotomy. But it must not be taken too far. For one thing, the *British* context was broader than the *English*, precisely because it included the Scottish universities, fast becoming both the key site of medical education and research, as well as the leading British center of moral and natural philosophy, of philosophical jurisprudence, and of systematic pedagogy [12]; [46].

Under such circumstances, it should not be at all surprising that the seminal compilation of medical ethics in the English tongue – John Gregory's *Lectures* – should have arisen out of a course at a Scottish university [31]. McCullough situates Gregory's system of medical ethics in the context of Scottish common sense philosophy, especially the thought of David Hume, the leading moralist of the Scottish Enlightenment [49]. The case, as McCullough shows, is actually less straightforward; for while Gregory himself was philosophically of the "common sense school," he was personally ill at ease with Humean skeptical views of religion; nonetheless (McCullough argues) although there are other philosophical conceptions of *sympathy*, those which underpin Gregory's medical ethics appear traceable to Hume.

In other words, as Baker and McCullough emphasize, British medical ethics had philosophical as well as practical roots, that need to be sorted out with care. Not least, as McCullough's account of Gregory, and Baker's analysis of Percival both stress, we must be eternally careful not to rewrite the meaning of classic texts of medical ethics after our own ahistorical prejudices and anachronistic preoccupations [49]. Baker argues that a number of the major commentators on Percival (specifically, Chauncey Leake, Jeffrey Berlant, and Ivan Waddington) have succumbed to precisely this temptation [6]; [47]; [82].

Reading the texts of Percival's code within the context of the problems facing eighteenth-century medicine in general, and the hospital environment in particular, Percival's codifications of medical morality are best interpreted, Baker argues, as marking a fundamental shift from the kind of individualistic codes of gentlemanly honor (discussed earlier by Fissell) to an intersubjective morality in which treatments, experiments, and disputes are determined and adjudicated intersubjectively, by practitioners acting, not as individuals, but in concert, as a collectivity. In Percival's ethic, Baker continues, medical practitioners exchange traditional moral autonomy – the right of each practitioner to assess his own actions by his own standards of medical and moral propriety – for collective standards of medical and moral propriety. What practitioners lost by forswearing their individual autonomy and becoming accountable to their peers, they gained back, however, by asserting a collective autonomy against their new patrons, the hospital trustees. This collective autonomy attendant upon collaborative self-regulation, Baker argues, formed the foundation of a sometimes implicit and sometimes explicit contractarianism which became the basis of the formal codifications of medical ethics in North America and Britain. In other words, the codified *collective autonomy* of practitioners was to become the moral basis of nineteenth-century medical professionalism.

Overall, this volume reflects those forces and fashions, which should promote understanding as to why it was in the latter part of the eighteenth century that medical ethics became a more important focus for speculation and encodement. In Britain, rapid commercialization was destabilizing the traditional professional hierarchy, creating opportunities for lucrative enrichment and dangers of competition and downgrading. New medical institutions, above all the hospital, were posing fresh problems regarding professional power, collective responsibility, and the division of labor. Gentlemanly codes of honor proved insufficient for the doctors, while the state, the administration, and the law courts offered few leads and little guidance. Under such circumstances, it was left to the medical cadres themselves to regulate their own enterprises and practices. They found, in the Scottish universities, for example, philosophical traditions upon which they could draw. There were evidently good reasons why the writings of Gregory and especially Percival proved valuable models for their successors. Above all, as Baker suggests, Percival's proposal of a self-regulating code ideally suited the circumstances of professional development within the sphere of *laissez-faire* capitalism in both Britain and the United States.

ROBERT BAKER
DOROTHY PORTER
ROY PORTER

REFERENCES

1. Ackerknecht, E. H.: 1964, 'Zur Geschichte der medizinischen Ethik', *Praxis* liii, 578–581.
2. Altschule, M. D.: 1975, 'What Medicine is About. Using its Past to Improve its Future', *Countway Library Associates Historical Publication*, 3, 100.
3. Amundsen, D. W.: 1977, 'Medical Deontology and Pestilential Disease in the Late Middle Ages', *Journal of the History of Medicine*, 32, 403–421.
4. Amundsen, D. W.: 1981, 'Casuistry and Professional Obligations; The Regulation of Physicians by the Court of Conscience in the Late Middle Ages', *Transactions and Studies, College of Physicians of Philadelphia* 3, 22–39.
5. Beddoes, T.: 1808, *A Letter to the Right Honourable Sir Joseph Banks ... on the Causes and Removal of the Prevailing Discontents, Imperfections, and Abuses, in Medicine*, London, Richard Phillips.
6. Berlant, J.: 1975, *Profession and Monopoly. A Study of Medicine in the United States and Great Britain*, Berkeley, University of California Press.
7. Boorstin, D.: 1973, *The Mysterious Science of the Law*, Gloucester, Massachusetts, Peter Smith.
8. Bulger, R. J. (ed.): 1973, *Hippocrates Revisited: A Search for Meaning*, New York, Medcom Press.
9. Burns, C.: 1977, *Legacies in Ethics and Medicine*, New York, Science History Publications.
10. Burns, C.: 1978, 'Medical Ethics, History of North America, Seventeenth to Nineteenth Century', in Reich, W. T. (ed.), *Encyclopedia Bioethics*, New York, Free Press, III, 963–968.
11. Burns, C.: 1974, 'Reciprocity in the Development of Anglo-American Medical Ethics, 1765–1865', *International Congress of the History of Medicine*, 23rd, London, 2–9 Sept. 1972; London, Wellcome Institute for the History of Medicine I, 813–819.
12. Camic, C.: 1983, *Experience and Enlightenment: Socialization for Cultural Change in Eighteenth-Century Scotland*, Chicago, University of Chicago Press; Edinburgh, Edinburgh University Press.
13. Carr-Saunders, A. M., Wilson, P. A.: 1933, *The Professions*, Oxford, Clarendon Press.
14. Childs, F.: 1984, 'Prescriptions for Manners in Eighteenth-Century Courtesy Literature', D. Phil. dissertation, Oxford University.
15. Clark-Kennedy, A. E.: 1969, *Man, Medicine and Morality*, Hampden, Connecticut, Archon Books.
16. Cook, H. J.: 1990, 'The Rose Case Reconsidered: Physicians, Apothecaries and the Law in Augustan England', *Journal of the History of Medicine.*
17. Cooper, R. N.: 1960, 'Medical Ethics Through the Ages', *Journal of Postgraduate Medicine* 6, 53–59.
18. Crawford, C.: 1987, 'The Emergence of Forensic Medicine: Medical Evidence in Common Law Courts, 1730–1830', Oxford, D. Phil.
19. Duncan, A. S. *et al.* (eds.): 1977, *Dictionary of Medical ethics*, London, Longmann & Todd.

20. Edelstein, L.: 1956, 'The Professional Ethics of the Greek Physician', *Bulletin of the History of Medicine* XXX, 391–419.
21. Edelstein, L.: 1967, *Ancient Medicine: Selected Papers of Ludwig Edelstein*, Baltimore, The Johns Hopkins Press.
22. Engelhardt, H. T., Spicker, S. F.: 1975, *Evaluation and Explanation in the Bio-Medical Sciences*, Dordrecht, D. Reidel.
23. Forbes, R.: 1955, 'A Historical Survey of Medical Ethics', *St. Bartholomew's Hospital Journal* 59, 282–286.
24. Freidson, E.: 1970, *Profession of Medicine: A Study of the Sociology of Applied Knowledge*, New York, Dodd, Mead.
25. Gelfand, M.: 1968, *Philosophy and Ethics of Medicine*, Edinburgh, E. & S. Livingstone.
26. Geyer-Kordesch, J.: 1985, 'The Cultural Habits of Illness: The Enlightened and the Pious in Eighteenth-Century Germany', in Porter, R. (ed.), *Patients and Practitioners*, Cambridge, Cambridge University Press, pp. 177–204.
27. Geyer-Kordesch, J.: 1985, 'German Medical Education in the Eighteenth Century: The Prussian Context and its Influence', in Bynum, W. F., Porter, R. (eds.), *William Hunter and the Eighteenth Century Medical World*, Cambridge, Cambridge University Press, pp. 177–205.
28. Geyer-Kordesch, J.: 1989, 'Passions and the Ghost in the Machine: Or What Not to Ask about Science in Seventeenth and Eighteenth-Century Germany', in French, R., Wear, A. (eds.), *The Medical Revolution of the Seventeenth Century*, Cambridge, Cambridge University Press, pp. 145–163.
29. Geyer-Kordesch, J.: 1990, 'Georg Ernst Stahl's Radical Pietist Medicine and its Influence in the German Enlightenment', in Cunningham, A., French, R. (eds.), *The Medical Enlightenment of the Eighteenth Century*, Cambridge, Cambridge University Press, pp. 67–88.
30. Gisborne, T.: 1794, *An Enquiry into the Duties of Men in the Higher and Middle Classes of Society in Great Britain, Resulting from their Respective Stations, Professions and Employments*, 2 vols., London, B. and J. White.
31. Gregory, J.: 1770, *Observations on the Duties of a Physician*, London, Strahan & Cadell.
32. Guthrie, D.: 1957, 'The Evolution of Medical Ethics', *Medical History* I, 70–100.
33. Hastings, J. *et al.* (eds.): 1908–21, *Encyclopaedia of Religion and Ethics*, Edinburgh, Clark.
34. Holmes, G.: 1982, *Augustan England: Professions, State and Society, 1680–1730*, London, George Allen & Unwin.
35. James, R.: 1743–1745, *A Medicinal Dictionary*, 3 vols., London, T. Osborne.
36. Jewson, N.: 1974, 'Medical Knowledge and the Patronage System in Eighteenth Century England', *Sociology* 8, 369–385.
37. Jewson, N.: 1976, 'The Disappearance of the Sick Man from Medical Cosmology, 1770–1870', *Sociology* 10, 225–244.
38. Johnson, T. J.: 1972, *Professions and Power*, London, Macmillan.
39. Karplus, H.: March 1972; 1973, 'Medical ethics in Paolo Zacchia's "Questiones medico-legales", in *idem, International symposium on Society, Medicine and*

Law, Amsterdam, New York, Elsevier Scientific Publications, pp. 125–134.
40. Katsareas, St. P., Costeas, F. X.: 1960, 'The Physician, a Cultured Gentleman', *Postgraduate Medicine* 18, 88–92.
41. Kepler, M. 0.: 1981, *Medical Stewardship, Fulfilling the Hippocratic Legacy*, Westport, Connecticut, Greenwood Press.
42. King, L. S.: 1958, 'Development of Medical Ethics', *New English Journal of Medicine* 258, 480–486.
43. Kolaja, J.: 1954, 'Historical Development of Medical Ethics in the United States', *World Medical Journal* 1, 155–157.
44. Kudlien, F.: 1970, 'Medical Ethics and Popular Ethics in Greece and Rome', *Clio Medica* 5, 91–121.
45. Larkey, S.: 1952, 'The Hippocratic Oath in the Elizabethan Period', *Bulletin of the History of Medicine* 26, 1–31.
46. Lawrence, C.: 1984, 'Medicine as Culture: Edinburgh and the Scottish Enlightenment', (Ph.D. dissertation) University of London.
47. Leake, C. D.: 1971, 'Percival's Medical Ethics, Promise and Problems', *California Medicine* 114, 68–70.
48. Loudon, I. S. L.: 1986, *Medical Care and the General Practitioner 1750–1850*, Oxford, Clarendon Press.
49. McCullough, L. E.: 1978, 'Historical Perspectives on the Ethical Dimensions of the Patient-Physician Relationship: The Medical Ethics of Dr John Gregory', *Ethics in Science and Medicine* 1, 47–53.
50. McDaniel, W. B. 2nd: 1956, *Thomas Percival's Privately printed "Medical Jurisprudence," 1794*, 36 Philadelphia.
51. McDonald, D. E.: 1962, *A History of American Medical Ethics, 1847–1912*, Madison, State Historicy Society for the University of Wisconsin.
52. MacKinney, L. C.: 1952, 'Medical Ethics and Etiquette in the Early Middle Ages: The Persistence of Hippocratic Ideals', *Bulletin of the History of Medicine* 26, 1–31.
53. McLean, S.: 1983, *Medicine, Morals and the Law*, Aldershot, Gower.
54. Pellegrino, E. D.: 1986, 'Percival's Medical Ethics': The Moral Philosophy of an 18th-century English Gentleman', *Archives of International Medicine* 146, 2265–2269.
55. Percival, T.: 1803, *Medical Ethics; or, A Code of Institutes and Precepts Adapted to the Professional Conduct of Physicians and Surgeons*, Manchester, J. Johnson & R. Bickerstaff.
56. Perkin, H.: 1969, *The Origins of Modern English Society 1780–1880*, London, Routledge and Kegan Paul.
57. Pickering, Sir G.: 1963, 'Manners Makyth Man. A Plea for the Importance of Character in Medicine', *Asian Medical Journal* 6, 23–28.
58. Pickstone, J. V., Butler, S. V.: 1984, 'The Politics of Medicine in Manchester, 1788–1792', *Medical History* 28, 227–49.
59. Porter, D., Porter, R.: 1989, *Patient's Progress: Doctors and Doctoring in Eighteenth Century England*, Cambridge, Polity Press.
60. Porter, R.: 1985, 'William Hunter: A Surgeon and a Gentleman', in Bynum, W. F., Porter, R. (eds.), *William Hunter and the Eighteenth-Century Medical World*,

Cambridge, Cambridge University Press, pp. 7–34.
61. Porter, R.: 1989, *Health for Sale: Quackery in England 1650–1850*, Manchester, Manchester University Press.
62. Porter, R.: 'Thomas Gisborne: Physicians, Christians, and Gentlemen', (Forthcoming) in Geyer-Kordesch, J., Wear, A. (eds.), *History of Medical Ethics.*
63. Probyn: 1974, 'Swift and the Physicians: Aspects of Satire and Status', *Medical History* 18, 242–256.
64. Quincy, J.: 1719, *Lexicon Physico-Medicum; or, A New Physical Dictionary*, London, A. Bell.
65. Radhill, S. X.: 1962, 'A History of Medical Ethics', *Philadelphia Medicine* 58, 873–876.
66. Ramsey, M.: 1982, 'The Repression of Unauthorized Medical Practice', *Eighteenth-Century Life* 7, 118–135.
67. Ramsey, M.: 1987, 'Property Rights and the Right to Health: The Regulation of Secret Remedies in France, 1789–1815', in Bynum, W. F., Porter, R. (eds.), *Medical Fringe and Medical Orthodoxy*, London, Croom Helm, pp. 79–105.
68. Ramsey, M.: 1988, Professional and Popular Medicine in *France, 1770–1830*, New York, Cambridge University Press.
69. Reich, W. T. (ed.): 1978, *Encyclopedia of Bioethics*, New York, Free Press.
70. Reiser, S. J. *et al.* (eds.): 1977, *Ethics in Medicine. Historical Perspectives and Contemporary Concerns*, Cambridge, Massachusetts, MIT Press.
71. Roberts, R. S.: 1978, 'Historical Aspects of Medical Ethics', *Central African Journal of Medicine*, 24, 214–216.
72. Shelp, E. E.: 1980, in Spicker, S. F., Healey, J. M., Engelhardt, H. T. (eds.), *The Law-Medicine Relation: A Philosophical Exploration*, Dordrecht, D. Reidel.
73. Shelp, E. E. (ed.): 1982, *Beneficence and Healthcare*, Dordrecht, D. Reidel.
74. Shelp, E. E. (ed.): 1983, *The Clinical Encounter. The Moral Fabric of the Patient-Physician Relationship*, Dordrecht, D. Reidel.
75. Bassford P., Bassford, S.: 1986, 'Codes of Medical Ethics: Traditional Foundations and Contemporary Practice', *Social Science and Medicine* 22, 1175–1179.
76. Spicker, S. F., Engelhardt, H. T.: 1975, *Philosophical Medical Ethics: Its Nature and Significance*, Dordrecht, D. Reidel.
77. Silvette, H.: 1967, *The Doctor on Stage: Medicine and Medical Men in Seventeenth Century England*, (ed.), Butler, F., Knoxville, Tennessee, University of Tennessee Press.
78. Szasz, T.: 1977, *The Theology of Medicine: The Political Philosophical Foundations of Medical Ethics*, Baton Rouge, Louisiana, Louisiana State University Press.
79. Veatch, R. M.: 1981, *A Theory of Medical Ethics*, New York, Basic Books.
80. Veith, I.: 1957, 'Medical Ethics Throughout the Ages', *Archives of Internal Medicine* 100, 504–512.
81. Veith, I.: 1959, 'On the Conduct of the Physician', *Modern Medicine* 27, 215–225.
82. Waddington, I.: 1975, 'The Development of Medical Ethics: A Sociological Analysis', *Medical History* 29, 36–51.
83. Waddington, I.: 1984, *The Medical Profession in the Industrial Revolution*,

Dublin, Gill & Macmillan.
84. Wear, A.: (forthcoming). 'The Early History of Medical Ethics', in J. Geyer-Kordesch and A. Wear, *et al.* (eds.) *History of Medical Ethics.*
85. Wolfenden, Baron, J. F.: 1978, 'Hunter, Hippocrates and Humanity', (Hunterian Anniversary Oration, 1978), *Annals of the Royal College of Surgeons* 9, 175–181.

PART ONE: PRE-HISTORY

MEDICAL PROPRIETY AND IMPROPRIETY IN THE ENGLISH-SPEAKING WORLD PRIOR TO THE FORMALIZATION OF MEDICAL ETHICS

INTRODUCTION

One of the classic thought experiments of philosophers from Plato to Hobbes to Locke to Rawls, is to imagine what the world would be like in the absence of ethics? What would our lives be like if there were no operant morality in terms of which people distinguished good from bad, right from wrong. In this section three historians, David Harley, Mary Fissell and Roy Porter describe the realities of such a world, the world of eighteenth century English medicine before the formalization of professional medical morality. They do so by acquainting us with the disputes and disputations of three eighteeth century practitioners: Thomas Beddoes, a Bristol physician (1760–1808); Henry Bracken, a Lancaster manmidwife, physician and surgeon (1697–1764), and Archibald Cleland of Bath, a hospital surgeon.

What is intriguing about the lives of these three eighteenth century practitioners is that, to an extent almost unimaginable today, their lives were enmired in disputes, not only in law courts and correspondence with other professionals, but in letters and pamphlets which they published to defend their personal and professional reputations. In the process they not only air the profession's dirty linen in public, but parade their own absence of standards to modern eyes. This is not to say that they have no sense whatsoever of wrongdoing. In a pamphlet countering charges by a Doctor Chistopherson that he was acting as a tradesman rather than a professional, Henry Bracken turns the charge against Christopherson, accusing him not only of misdiagnosing and misprescribing but of making "the noble Art of Healing more of a Trade than a Science, and consequently you prescribe more for your own and the Apothecary's gain than the Patient's health, otherwise you would have no occasion to dangle so much after people in imaginery illness." As charge and countercharge

R. Baker, Dorothy Porter and Roy Porter (eds), The Codification of Medical Morality, 15–17.

make clear it was considered as improper then, as it is today, for a physician to prescribe for personal profit. What was unclear, as is evident from Bracken's disputes over non-surgical cures for "the stone" and from Thomas Beddoes' diatribes against the profession's "lukewarmness towards improvement, which so commonly arises when the physician has gotten full scent of profit," was how to distinguish the genuine article from the imitation, the true physician from the quack, how to give the practitioner the science and the integrity to resist not only avarice but also the "pressure to prescribe for the alarms of the patient."

Neither Beddoes nor Bracken knew how to prevent Hygeia from becoming slave to Plutus, but they, along with their contemporaries would tend to agree that at the root of all ethical medicine was the distinction between medicine as an art in service of humanity, and medicine as a commercial endeavor, engaged in primarily for the profit of its practitioners. Beddoes thought that recruiting properly educated gentry into the profession would prevent the corruption of Hygeia by Plutus and Bracken claimed to be personally non-corrupt by virtue of his education, his commitment to scientific medicine, and his willingness to admit and correct earlier errors in his practice. But, and here is a case in which absence is significant, while they both agree on the importance of elevated motives and scientific practice, neither Beddoes, in stating his program for the reform of medicine, nor Bracken, in his various ripostes and defences ever mentions, much less alludes to, the classic formulation of these ideals in the Hippocratic Oath.

The dominant myth in the history of medical ethics is that of the Hippocratic footnote, the idea that the foundations of Western medical ethics were laid down in the Hippocratic Oath, and/or the accompanying corpus, and that the history of medical ethics from that time to the present is essentially a footnote, or, if not a footnote, at least a series of comments and reflections on premises laid down in the Oath. It is difficult to reconcile this view with what we know of eighteenth century medicine, especially as it surfaces in the writings and disputes of Beddoes and Bracken. It is even more difficult, as Mary Fissell points out in her study, to reconcile the purported dominance of Hippocratic morality with the absence of any specific mention the Oath or the aphorisms in the Cleland case; for at issue in that case was an act which clearly violates one of the classic Hippocractic prohibitions, the prohibition against having sexual intercourse with one's patient.

On her death bed a woman accused the hospital's surgeon – he had done her Business – and the Board of Governors of the hospital took her charge seriously. Yet, as Fissell's analysis shows, no party to this dispute parses the issues in Hippocratic terms. The Governors never accuse Cleland of violating his Oath, but rather of acting "contrary to all the rules of Christian Charity"; they point out that the "objects of a HOSPITAL CHARITY are *Helpless* and liable to every kind of *ill treatment*; if they are not protected by those, to whose Care they are entrusted." Cleland defends himself by arguing that vaginal examinations are standard medical practice, and that "no man would have thrust his Fingers into the *Common-shores* of a couple of *Nasty pocky leperous Whores*, but out of Laudable Zeal of being helpful." As Fissell notes, the underlying conceptual framework here is not physician and patient, but donor and recipient. Moreover, not even Cleland takes refuge in the Hippocratic ideal of asexual touching, indeed, he does not even characterize these women as patients, but as "leperous whores," and the recipients of the Hospital's charity.

What emerges from the cases examined in this section is a picture of medicine prior to the emergence of accepted conventions of professional morality, a medicine in which neither practitioners nor patients nor patrons accepted that medicine, as a profession, was to be judged by a special medical ethic. The disputes, incidents and reforms described here, therefore, can be said to belong, in a sense, to the pre-history of modern medical ethics.

ROBERT BAKER

CHAPTER 1

MARY E. FISSELL

INNOCENT AND HONORABLE BRIBES: MEDICAL MANNERS IN EIGHTEENTH-CENTURY BRITAIN

While the shade of Hippocrates looms large in our current assumptions about the roots of medical ethics, early modern medical practitioners rarely looked back to antiquity for guidance about ethics. Indeed, no ethics particular to their profession or vocation governed conduct. Rather, appropriate behavior was inculcated through the institution of apprenticeship, shaped by general norms of master/servant and client/patron interactions. It was only in the 1770s that a medical ethics became possible or desirable, following changes in the structure of medical practice and shifts in more general cultural assumptions about behavior.

I. MANNERS AND MORALS

In the late seventeenth and most of the eighteenth centuries, the behavior of medical practitioners – surgeons, apothecaries and even physicians – was governed by general codes of conduct, by the norms and constraints described by "manners" or "courtesy." But "manners" implied far more than the limited version of etiquette that the term means today. The rich and extensive courtesy literature of the renaissance, such as Thomas Elyot's *Book of the Governor*, addressed the whole realm of social relations and the nature of society itself. This combination of prescriptive and analytical elements continued to be a part of eighteenth-century texts on manners. For example, Nathaniel Lancaster's *Essay on Delicacy* (1748) was to have sections on the origin of society and its laws, as well as specific details of politeness in writing, speaking and painting, and examples of good conduct [8], [12], [22], [39], [49].

In other words, manners both prescribed correct behavior in certain settings and embodied particular ideas about how people lived in groups, how social structures functioned, and how individual conduct and society overall were connected. Originally, this literature was directed towards an aristocratic audience; one can probably consider Castiglione's *Advice to a*

R. Baker, Dorothy Porter and Roy Porter (eds), The Codification of Medical Morality, 19–45.

Courtier as the *ur*-text of this type. But by the mid-seventeenth century, such courtesy books were read by a broader segment of the population, eager for glimpses of high life and ready to emulate their betters. So too, writers began to produce texts for this audience, aimed at middling sorts rather than the scions of the aristocracy. The 1730s, for instance, saw the publication of *The Man of Manners: or Plebian Polish'd. Being Plain and Familiar Rules for a modest and genteel behavior, on most of the ordinary occasions of life.*

Obviously, these discourses were a function of a literate print culture. In linking the precepts of such texts to the norms of surgeons' and apothecaries' practices, we move into a culture which often relied upon oral rather than written sanctions. Apprenticeship remained a face-to-face training predicated on oral transmission which remains largely impervious to historical eavesdropping [31], [32], [51]. So too, outside the metropolis, city companies – the regulatory bodies of provincial medicine – conducted their business in courts and assemblies which did not always require or generate written records. While apothecaries and surgeons were undoubtedly literate, integrating printed discourses written for a variety of purposes with the actual functioning of behavioral sanctions remains problematic for the historian

However, comparing texts which purported to instruct medical youths with those intended for a more general audience suggests that advice to medical apprentices derived from the more general conduct literature. To put it the other way, medical men were not presumed to rely upon specifically medical codes of behavior. Both general works of advice on manners for the middling sorts and those few works directly addressed to the medical man have certain key precepts of behavior in common, which point to a lack of professional identity and ethics and towards a more widely-shared code of conduct.

Virtually everyone who wrote about education for the professions during the eighteenth century agreed that manners mattered. James Nelson, an apothecary who wrote a book on the education, manners, and health of children in 1753, said "all learning which does not improve our Manners, is vain and unprofitable" ([40] p. 23). James Lucas, author of a book on the education and role of the surgeon-apothecary, enjoined that, "a surgeon should not only possess professional qualifications, but these should be constantly adorned with virtuous principles and engaging manners" ([36], p. 76). William Chamberlaine, a surgeon-apothecary, faulted "unmannerly" behavior in an apprentice ([11], p. 37). Stephen

Philpot, a dancing master, wrote that a highly-qualified man could be "quite over-looked and disregarded, and for no other Reason but the want of that genteel Address and Deportment which are necessary to recommend him in the World" ([43], p. 54). In other words, manners were an essential component of a medical persona.

Two kinds of social relationships provided models for medical apprentices: master/servant and client/patron. Neither was peculiar to medicine. Both served to illustrate how people of different ranks could deal with each other. William Chamberlaine's guide for the conduct of apprentices, for example, repeatedly justified precepts of behavior by linking them to general expectations of master/servant or client/patron interactions. Although published in 1812, Chamberlaine had written the manuscript almost 30 years earlier, for the guidance of his apprentice, and the book provides a rare glimpse of the day-to-day routines of a surgeon-apothecary and his shop in the mid- and late- eighteenth century.

For instance, Chamberlaine discusses the apprentice's conduct in his Master's (always capitalized) home. After dinner, when the cloth is taken off the table, the apprentice should always leave, since whatever his family's rank, he is of lower status than his employer. Chamberlaine explains this precept in terms of wider social norms,

> ... it is contrary I believe to the usual and established custom, in all trades, and professions, for assistants, shopmen, or apprentices to remain in the room after the cloth is removed ([11], p. 147).

He adds that "if you have been a good lad, and have always conducted yourself well, modestly, and as a gentleman," then the master will likely invite the apprentice to stay for special occasions. But even in this instance, the apprentice must display appropriate deference, offering to leave after the first glass of wine. Only if asked to remain again may he stay and enjoy himself ([11], pp. 148–149). Such precepts pointed to more general notions about how human society worked, such behaviors in this case illustrating that both master and apprentice appreciated the difference in status between them.

But it was not just in behavior towards one's betters that the master/servant paradigm functioned. Courtesy books dealt also with the delicate area of conduct towards those of inferior status. Most young men were trained in households which were themselves carefully graded hierarchies of family members, apprentices, and servants. There were niceties of distinction even amongst the apprentices. Custom permitted senior apprentices, for instance, to wear their hats inside the house and even inside the shop, and these privileges were carefully guarded ([21], pp. 102, 226).

And, of course, once the apprentice had grown up and become a master in his own right, the same kinds of strictures bound him in relation to his own apprentices. It was up to the master to maintain correct and deferential relations with his apprentices. As Daniel Defoe cautioned, "Easy masters make sawcy servants" ([19], p. 261). Young's chronicle of the London company of barber-surgeons includes the instance of two young men failing their examinations because of "sauciness" towards their examiners ([55], p. 349). William Chamberlaine, whose text recounts numerous instances of apprentice peculation, laziness, and ineptitude, was most exercised by sauciness. In his list of queries for a boy's former employer, "DOES HE GIVE SAUCY ANSWERS?" is the only item in capital letters. For Chamberlaine, and others like him, suitable courtesy to one's master was the keystone of a successful master/servant relationship. Apprenticeship was intended to maintain appropriate and deferential social relationships, not promote quick-tongued witty apprentices.

The emphasis placed upon social relations in manners books meant that these kinds of hierarchical structures were particularly evident in conduct. Thus, for instance, William Chamberlaine slid easily between manners, morals, and specific behaviors in his advice to masters seeking apprentices:

> Let the master enquire most particularly into the morals of the youth he is about to take *as a son*, for six years. Whether he is cleanly in his person and dress; affable and gentlemen-like in his behavior to customers, and all persons coming on business; honest in money matters; ... whether fond of the kitchen, the company of servants, and of low company in general ([11], pp. 10, 11).

Moral worth could be assessed by courtesy, and a master overlooked bad manners at his peril.

While the potential contradictions between appearance and reality masked by courtesy were occasionally alluded to by some writers, far more common were those who instructed that manners and morals went together, that appearance and reality were matched. As an anonymous dialogue of 1715 had it, a medical man "ought to make his Appearance and Conversation as Agreeable, Genteel, and significant as may be" ([2], p. 95). Much of the concern about appearance was related to considerations of the client/patron relationship.

William Chamberlaine spent several pages of his advice book detailing how apprentices should interact with servants, especially the maidservants who came to the shop for their mistresses' medicines. Here both apprentice and maidservant were standing in for their employers, each taking the client or patron role which the patient/doctor encounter predicated. He said, "The best way of gaining the respect of this class is by a dignified civility, and never forgetting the respect that is due to yourself" ([11], p. 158). In other words, although a client, one should not over-humble oneself. He counselled the apprentice not to be too witty, and not to tell jokes to maidservants, because there were those who "misconstrue the most harmless expression into a lessening of their importance, and return home to their mistress with an avowal that they will never go to that shop again" ([11], p. 158).

Similarly, Chamberlaine discussed how to behave towards one's patron/patient when blood-letting. This was especially important advice for apprentices, since busy masters often permitted their apprentices to bleed patients and sometimes to keep the fee for themselves. An inept apprentice could thus lose the master an important patient. Chamberlaine counselled the apprentice to pay especial attention to post-blood-letting care. For male patients, it was important to roll down their sleeves after attending to the bandage. For women, it was essential to ensure that no trace of blood remained on their arms,

> Such little attentions gain you respect, and impress on the minds of strangers, that you know your business, and good manners ([11], p. 156).

One of the few occasions on which a practitioner might routinely touch a patient was thus transformed into a careful ritual, one which attempted to compensate for the transgressive nature of the encounter. The blood-letter's courteous attention to returning the patient to his or her untouched status underlines the mixture of courtesy and technique which made good medical practice.

Much of the skill of the client in pleasing the patron lay in his manners, or address. Chamberlaine, in his advice to masters seeking medical apprentices, told them to notice "whether he is dirty in his person, or morose in his address" ([11], p. 12). As the word "address" suggests, an important part of this relationship depended upon personal appearance [12]. Medical men were supposed to look like gentlemen. As a 1715 dialogue on the duties of medical men had it, a doctor "should go Neat and elegant in his Dress" ([2], p. 95).

James Nelson, author of a text on the education of children, recounted an episode which clarified the function of medical attire. In an era none too distant from sumptuary laws, dress was intended as a revelation of social status. Nelson described a visit to a friend of his, an apothecary:

> ... a Young Fellow, my Friend's Apprentice, was at Work behind the counter, and out peep'd a lac'd Waistcoat (I must observe it was in the Days when a lac'd Waistcoat stood for something, for it has now, I think, lost all its Significance).

Nelson, startled to see such attire on an apprentice, asked who the lad was, and was told that he was the son of a coachman. The lace on the waistcoat was from his father's livery. Nelson scolded, "Thus what was before no better than the Badge of Dependence, is now turned into an instrument of Contempt and Ridicule" ([40], p. 373). The critique of insincerity implied in Nelson's comment was to become a dominant theme later in the century, but in the 1750s, Nelson could still expect that dress accurately reflected rank.

Wigs also functioned as badges of status. James Collyer, author of *The Parents and Guardians Directory*, 1761, had this to say:

> tis certain that if a young Doctor should venture to step out of the common road, and wear his own hair, or a campaign or bag wig, he would find it hard to convince the world that he was either wise or learned.

Although Collyer added that, "A large periwig can be of little help to the intellect of the head it covers," aspirant medical men should adopt the mien and appearance of their fellows ([16], p. 43). James Lucas counselled that "sedateness of Character being requisite should prohibit fantastical Apparel" ([36], p. 3). Of course, display and personal adornment were also a means of advertisement or at least of distinguishing oneself from the rest of medical practice; some balance between decorum and distinction had to be struck.

But it was not just apparel that marked out the gentleman from the clod. In ways now unfamiliar to us, deportment and the way in which one used one's body provided important social cues, what Erving Goffman refers to as kinesics [25]. As Collyer said about a would-be apothecary:

> He should have a genteel person and behavior; for one who has naturally the clumsiness, the walk, the air, or the blunt rudeness of a plowman, can never be fit for this genteel profession ([16], p. 45).

An essay from 1715 was more specific about how one was to carry one's body:

> His Gate and Motion should be Uniform and Equal. He should neither Stalk nor Hurry into a Chamber. To come in Gasping, and Staring, and Writhing his Mouth, shews a Man wants Recollection and Presence of Mind extreamly ([2], p. 102).

These only echoed the precepts of other experts on manners and decorum far beyond the world of medicine. Nor was such guidance only for gentlemen. *The Man of Manners: or Plebian Polish'd* offered the same advice on deportment, "If whilst we are walking, we see any person of our Acquaintance ... we are not to *Bawl* and *Hem* after them, like a Butcher out of a Tavern Window" [7]. The appearance of gentility was its possession; good manners lacked the implication of deception and display that they were later to gain. Thus, for example, the 1715 *Essay on Duty* was able to say, "In short, good House-keeping and Elegancy of Living is an Innocent and an Honorable Bribe" ([2], p. 99).

These recondite skills of "address" and "manners" were important if a young practitioner were to make a career for himself. In the words of James Lucas:

> Affability, and polite manners form a professional man for an easy admittance into the company of his superiors, and afford him frequent opportunities of gaining their esteem ([36], p. 80).

Again and again, writers emphasized the importance of being able to mix with one's betters, an ability dependant upon familiarity with ceremony and form, upon an easy civility. The very lack of rigid social structures demanded that demarcations of rank could be read through manners, as James Nelson explained:

> In a Society of Men, suppose a Coffee-House, we see a promiscuous Crowd of Gentlemen and Tradesmen; in an assembly of Women, we see mix'd with the Gentry, not only the Wife of the Merchant, but that of the Brewer, the Distiller, the Druggist, and the Draper; and it is highly necessary that these should have such education, and their Manners so regulated, as will make them fit Company .... ([40], p. 359).

In the fluid urban social world of the eighteenth century, society itself could only function if people shared an understanding of a differentiated set of courteous behaviors.

In sum, the traditional association between manners and morals, courtesy and virtue, provided a means of discussing behavior towards one's betters and inferiors well-suited to the educational process of early modern medical practitioners. The structures of apprenticeship and household-based professional service were identical to those in other types of work, and advice on manners found many an audience. The key text for early modern medical ethics lies, not in Hippocrates' *Decorum*, but rather in the many descendants of Elyot's *Booke of the Governor*.

II. MEDICAL MANNERS AND ARCHIBALD CLELAND

However, codes of conduct inculcated through apprenticeship and reinforced by city companies were not foolproof. Ironic commentators noted the potential for insincerity in manners which might mask dubious morals. Nor was behavior unequivocal. An incident in Bath in the 1740s illustrates how complex issues surrounding manners and morals could become, particularly amongst medical men. In this case, hospital governors and a hospital surgeon mobilized an array of definitions of correct behavior in their dispute, but neither side grounded their claims in a peculiarly medical code of conduct.

Archibald Cleland was born in Scotland, apprenticed to an Edinburgh surgeon, and then came to London in 1720 to pursue his education at St. Thomas' and thence to Paris. He served as a ship's surgeon with the South Sea Company, took up private practice in Jamaica, and when he returned to England in the early 1730s, became a regimental surgeon. In 1741 he set up practice in Bath, and when the General Hospital opened in 1742, he applied for a post, and with the backing of the Earl of Chesterfield and Alexander Pope, became an assistant surgeon [14], [47].

But within the year controversy erupted. Cleland had examined three women, Sarah Appleby, Mary Hooke, and Mary Hudson, in the late summer and early autumn of 1743, some on more than one occasion. Appleby had applied to Cleland, evidently for charitable medical care, long before she was admitted to the Bath Hospital suffering from the aftereffects of a miscarriage brought on by venereal disease. (It was not clear to the medical staff that she was suffering from venereal disease or she would not have been admitted.) Dr. Rayner, one of the infirmary's physicians, under whose care Appleby remained, asked Cleland to perform a uterine injection since she seemed to be discharging bits of placental material ([14], pp. 18–20; [4], pp. 3–6). Cleland complied, although he was not the surgeon assigned to Appleby. Then Mary Hooke asked to be examined; it is not clear from the record why she requested this, and whether she expected a vaginal examination or merely a routine looking-over. Although Hooke was not Cleland's patient either, he claimed that he thought she was pregnant, and thus deceiving the charitable institution, so he examined her. Mary Hudson's case was a bit more straightforward; she was suffering from fits, after a miscarriage, and Cleland examined her because he thought her illness was the same as Sarah Appleby's ([14], pp. 22, 34).

Cleland was called up before the staff of the Infirmary to explain his actions: he had erred, not only in the type of examination of these women, but also in examining patients who were assigned to the other surgeon. Subsequently, at an extraordinary meeting of the hospital governors, Cleland was dismissed. However, relations between professional men – which were later to be a focus of medical ethics – were not central to this dispute. There was little discussion of the issue of whose patient was whose; rather, issues surrounding the actual examination were argued and re-argued.

The three women all objected to Cleland's examinations (although at what point they voiced complaints is unclear). Mary Hooke and Mary Hudson both testified that his examinations had hurt them and caused them to bleed. Hooke additionally claimed she was examined in the nurse's room, with the door bolted, against her will. The most dangerous testimony was secondhand; Mary Hooke recounted what she'd been told by Sarah Appleby the night that Appleby died:

> ... [Appleby] begged her, for God and Christ's Sake, that he [Cleland] might never touch her [Hooke] no more, for that he wou'd bring her into the same way; and wish'd she had never been under his Care, and that he had done her Business ([4], p. 6).

'Business' meant sexual intercourse in contemporary slang, and soon the rumor spread that Cleland had been dismissed because he had debauched two women at the hospital ([15], p. 45).

In the ensuing pamphlet war between Cleland and the hospital governors, four themes were repeatedly contested, with various protagonists arguing for different interpretations of these four areas. One crucial theme is that of charity: What are charitable motives? How can we know them? What is cheating a charity? Second is the medical terrain, that least favored by any of the opponents. Here arguments were mobilized concerning the physical states of the women. But these arguments repeatedly shaded into those about character. Two final areas of debate concerned the characters of the women examined and those of the disputants. But none of these contests centered on behavior appropriate to a medical man or to a medical institution. Instead, all disputants played upon more widely held concepts of suitable conduct [29].

Indeed, at the meeting of the committee of hospital governors, it was originally suggested that Cleland's actions might have been criminal. Cleland tried to deflect this suggestion by returning the debate to the question of his propriety in examining another surgeon's patients, but the committee ignored his attempted diversion. The criminal nature of Cleland's act was never spelled out: Rape or assault would have been very difficult to prove [13]. The committee seems to have changed the charge against Cleland to one of misbehavior just prior to the vote in order to muster greater support for the anti-Cleland faction. However, the potential construction of Cleland's actions as criminal again points to the committee's refusal to construe the event as primarily a medical one.

Both Cleland and the governors tried to define their behaviors as appropriate to an act of charity. For example, in Cleland's first pamphlet, he argued that he should have been offered the chance to refute the women's testimony against him; not being afforded this opportunity was "entirely subversive of all the Rules of Charity and Civil Society" ([14], p. 8). Later in the same text he claimed that "Charity and the Desire of Relieving the Afflicted" were his only motives for examining the women ([14], p. 21). In a subsequent salvo, he argued that the governors were themselves lacking in charity, judging him, "contrary to all the rules of Christian Charity" ([15], p. 28).

The governors, however, seem to have been more successful in their utilization of concepts of charity against Cleland. They claimed that he had acted improperly within a charitable institution:

> The *Rich* and *Powerful* are capable of *repelling Insults* and *Punishing Injuries*; but the *Objects* of a HOSPITAL CHARITY are *Helpless* and liable to every kind of *ill Treatment*; if they are not protected by those, to whose Care they are entrusted ([4], p. 31–32).

Here the women whom Cleland examined are referred to as Objects of Charity rather than as patients. Nowhere in the accusations and counter-accusations do we see this incident in light of the Hippocratic injunction which forbids sexual intercourse between doctor and patient. The absence of such references underline the point that what is at issue is not medical ethics (as we would understand this subject today), but correct behavior (etiquette) between those of different social ranks – one of the key topics addressed by courtesy literature.

The multiple meanings of the term 'charity' in this dispute help to explain how and why both sides could claim to be in the right. Cleland often used the word in its sense of giving someone else the benefit of the doubt, a sort of courtesy. Thus he can equate the rules of "charity" with those of "Civil Society," and accuse the governors of lacking charity. But the governors focused on a newer meaning of the word, using it metonymically for a charitable institution. Thus they stressed the eleemosynary aspects of the setting in which Cleland worked, an aspect to which he alluded only once.

Cleland was unable to change the terms of this debate by medicalizing these women by turning them into "patients"; the governors' use of "charity" prevailed. As we shall see, discussions of these women's medical status inexorably slid towards their moral status, in both Cleland's and the governors' accounts. What Cleland could do was to utilize the governors' definition of the argument as one concerning charity, and to claim that these women were defrauding a charitable institution. For example, he maintained that Mary Hooke was pregnant, and hence ineligible for the charity's help. He examined her in order to prevent the hospital "from being imposed upon" ([14], p. 21).

The key to the dispute lay in character. When either side attempted to use the rationale of medicine or science, the issue of character took over. For example, Dr. Oliver, the chief physician at the hospital, and the leader of a segment of Bath's medical men, attempted to use the evidence of a post-mortem to discredit Cleland. According to Cleland, Oliver suggested that Cleland's examinations of Sarah Appleby and the injections with which he treated her were inappropriate, because at post-mortem "the Uterus appear'd to him in a Virgin State," and thus she could not have been suffering the aftereffects of a miscarriage. Cleland delighted in recounting that the rest of the faculty had to inform Oliver that Appleby had had an illegitimate child and was suffering from a miscarriage induced by taking medicines improperly. But Oliver's inability to construct a plausible account of Appleby's illness in no way impugned his medical authority in Bath, although Cleland used it to good textual effect.

In a similar fashion, Cleland claimed in his second pamphlet that vaginal examinations were approved of by such medical authorities as Riverius, Bartholinus, and Diociis. But again, this recourse to medical authority was just a jumping-off point for Cleland to attack the characters of his opponents, whom he argues were "void of shame" in making their suggestions ([15], p. 58). Nor does Cleland explicitly argue that the governors were void of shame because they overlooked medical authority; again, this accusation could be read on multiple levels. Rather, he inverts the criticisms of his character by applying them to the governors.

Much of this dispute was constructed around issues of behavior, credit, and reputation. Cleland cast aspersions on the characters of the women he examined and the governors who found him guilty of indecent practices; the governors claimed to be able to read Cleland's character and find it wanting. Certainly Cleland's position as a newcomer to Bath, who had dared to publicly criticize the management of the baths, was central to this dispute between insiders and outsiders ([52], pp. 32–38). So too, debates about the propriety of man-midwifery no doubt provided a subtext which contributed that frisson of interest to this controversy [45]. But the crucial issue was not about the propriety of doctors examining patients or men touching women – it was about character and behavior in a charitable institution.

Cleland, whose pamphlet was published first, established the terms of discourse in his introduction, pointing out that "our Credit is undeservedly stained, and our Reputation unjustly blasted" ([14], p. i). The governors, when they published their vindication, also cited character as a key issue:

> The reader is now desired to consider the *private Characters* of the Thirteen Gentlemen who voted for *Mr. Cleland's Dismission*. They modestly hope, that, upon the severest Scrutiny, they will appear upon the Level with the Rest of Mankind, who enjoy a *fair Reputation* ([4], p. 5).

In other words, the reader was to judge between Cleland and the governors based upon character, just as a prospective patient might choose between practitioners. Cleland also employed the rhetorical appeal to the reader, asking him or her to judge both the characters of the governors and, by implication, that of himself ([15], p. 58):

> is it possible to believe, that *Thirteen Gentlemen could be so void of all Shame*, as to impute this Enquiry [ie., the examinations] to a Desire of gratifying a *vicious beastly Inclination*?

In other words, if the governors were the gentlemen that they claimed to be, they could not lower themselves to think that Cleland could have loathsome motives.

What made these accusations and counter-accusations more powerful was the way in which the disputes about the characters of the women seemed to parallel and mimic those about Cleland and the governors.

Cleland claimed that Mary Hudson,

was a *cheat*, an *Impostor*, a great *Liar*, *had miscarried by taking of medicines*, and that *no Credit* ought to be given to any Thing she could say ([14], p. 33).

Cleland, faced with an accusation of sexual misconduct, stood the argument on its head and made the women's testimony invalid by suggesting that *they* were guilty of sexual misbehavior. He found a witness to suggest that Mary Hooke had falsely sworn a rape against a gentleman at Cheltenham, and was thus doubly untrustworthy ([14], p. 28).

But the hospital governors had an answer to that: "is a woman who wants Chastity incapable of swearing the *Truth*" ([4], p. 21). In effect, the governors beat Cleland at his own game by dissociating sexual reputation, honesty, and worth. As Susan Amussen has shown, the word 'credit' which appears again and again in this dispute held a double meaning in early modern England. It referred both to the probity of a witness in court, and to the financial stability of the individual, linked through the presumed refusal of bribes for testimony. A person of "no credit" implied both the potential for dishonesty and the more absolute lack of economic resources ([1], pp. 151–155). As wealth became associated with moral worth, it is easy to see how poor Mary Hooke could be portrayed as lacking credit in every sense of the word. Indeed, Cleland was accused of bribing Mary Hooke's mother in an attempt to get her to persuade her daughter to change her story – which, of course, paradoxically lowered his credit.

Character was central to this dispute because Cleland himself admitted that the question of impropriety hinged upon his own intentions, and the governors noted that he "allowed many of the facts contained in the Affidavits to be true" ([4], p. 9). Had Cleland "vicious" or "beastly" intentions, then his act would be criminal ([15], p. 58; [3], p. 2). Cleland indulged in misogynist purple prose in order to shift the "beastly" qualities from himself to the women:

no Man living would have thrust his Fingers into the *Common-shores* of a Couple of *Nasty pocky leperous Whores*, but out of a Laudable Zeal of being helpful ... ([15], p. 58).

Cleland also used the trope of anti-Popery to suggest that the governors could not know his motives:

As to their charging me with having acted in the Examinations of these Women from *vicious motives*: I say, this ... is to act upon the detestable Principles of the Inquisition ... ([15], p. 28).

However, the governors claimed that they could read character, that Cleland was indeed guilty of "indecent practices" ([4], p. 18).

The Cleland affair illustrates how issues of manners and morals could be contested when participants did not share the same interpretations of behavior. For both Cleland and his opponents, Cleland's actions were potentially "vicious." What determined their moral content rested upon character, which could be read through manners. Cleland's occasional attempts to use the authority of medicine were largely unsuccessful – even he himself could not seem to push them to their conclusion, shifting instead to issues of character. Even more striking is the complete absence of any mention of medical ethics. Neither Cleland nor his opponents constructed their arguments around a notion of medical behavior or medical institutions. Rather, the peculiar status of the object of charity and the character of Cleland determined the rightness or wrongness of the incident.

III. THE CRITIQUE OF MANNERS AS MORALS

By the later part of the eighteenth century, medical manners and morals became unglued; no longer were codes of conduct based on courtesy functional. Precepts that were to be taken seriously in 1700 were the stuff of fun by 1800. This crisis was due in part to the overall decline in the significance of manners as a discourse on social relations. As Michael Curtin has delineated, manners disappeared as a form of advice literature somewhere in the late eighteenth century, only to be replaced in the 1830s by the much more narrowly conceived etiquette book. Curtin lays the blame partly on the Earl of Chesterfield's shoulders.

It was Chesterfield's letters to his son, published in 1774, which revealed what many already knew and accepted – that good manners were not so much the sign of innate virtue as the indicator of social expediency [46]. Certainly Chesterfield was not the first to note that good manners had their uses. Bernard de Mandeville characteristically analyzed politeness as benefitting the individual by making society bearable in his *Fable of the Bees* in 1714 [18]. But Chesterfield went further. The truly courteous and successfully polite individual "knew the various workings of the heart, and artifices of the head" and thus could "employ all the several means of persuading and engaging the heart." This manipulative quality of politeness was denounced by others and seems to have been the element of Chesterfield's work which has led historians to pinpoint it as leading to the more general discrediting of manners literature.

But three other factors contributed to the death of an ethic based on manners. Related to the post-Chesterfield demise of manners was the critique of insincerity generated by certain Scottish Enlightenment thinkers, linked to the creation of the literary "man of feeling." However, it was not just in the realm of ideas that manners lost their allure; the commercialization of courtesy spelled the end of its moral power. So too, changing contexts of practice distanced medicine from the trade-based ethic of the shop and altered the client/patron structure of patient/doctor interactions. While some of the connections remain tentative, the death of manners as medical ethics in the 1770s was clearly overdetermined.

One attack on the insincerity of manners derived from the emergence of the man of feeling, that literary and prescriptive character who embodied the opposite of Chesterfieldian coolness by bursting into tears every ten pages or so, emphasizing the delicacy of his emotional makeup and the lack of artifice that went with it. The man of feeling has engendered many pages of analysis in literary circles; what is significant to this discussion are the Scottish Enlightenment roots of Henry MacKenzie's *The Man of Feeling* and its two best-selling successors in the 1770s. Richard Sher, Nicholas Phillipson, John Dwyer and others have discussed the quest for civic virtue in a commercial polity which absorbed the cultural energies of the Scottish Enlightenment [20], [30], [42], [44], [50]. The message of the Scottish ethicists and the man of feeling echoed the critiques of Lord Chesterfield. While the delicacy of manners of the modern era pointed to the civilized refinement which the age had achieved, its by-product was a manners separated from morals and subject to the worst kinds of insincerity and dissimulation ([20], pp. 60–61).

For example, in 1775, an anonymous "Young Gentleman" published one the first post-Chesterfield works on manners. His disdain for artifice echoes, not only Chesterfield, but also – perhaps unwittingly – the Scottish ethicists' championing of sincerity. He disdained appearance,

> The qualifications and marks of a real gentleman do not consist in elegant mansions, rural villas, spacious parks, treasures of gold, or a numerous retinue of servile attendants – these are only the appendages of grandeur, pomp, and parade; but propriety in thinking, justice in acting, and generosity in feeling are the figures by

which the gentlemen is to be known ([5], p. 28).

This emphasis upon sincerity rather than manners was echoed in the medical world by John Gregory, the Edinburgh professor who translated the Enlightenment critique of manners into medical precepts.

Gregory's lectures on the duties and offices of the physician and his *Father's Legacy to His Daughters* both address the social function of manners and their relation to ethics. Both criticize the artificial quality of contemporary manners, and ground that critique in a more fundamental dislike of luxury and dissipation. But manners were also necessary to social life, and hence Gregory could simultaneously fault insincere manners while promoting good ones. He also represents a departure from earlier English language writers, in that he proposes an ethic particular to medicine.

In the essay addressed to his daughters, Gregory continually stresses the importance of sincerity and the evils of artifice. He faults worldly manners as artificial, preferring that his daughters, "possess the most perfect simplicity of heart and manners" ([28], p. 45). Genuine sentiment was always contrasted to "The luxury and dissipation that prevails in genteel life" ([28], p. 63). As for medical men, so too for women: There are rules of conduct specific to their particular role, which are integrated with more general precepts about behavior ([28], p. 7).

In his medical lectures, Gregory plays upon the same themes of insincerity and genuine sentiment. At the beginning of the book, he prefigures much of his argument by setting up a comparison of two kinds of medical men:

> ... some have acted with candour, with honour, with the ingenuous and liberal manners of gentlemen. Conscious of their own worth, they disdained all artificial colourings, and depended for success on their real merit ([27], p. 3).

Such paragons of virtue, however, were not common and hence, "we never meet a physician in a dramatick representation, but he is treated as a solemn coxcomb and a fool" ([27], p. 4). This passage employs all of the key words of Gregory's analysis: good physicians can be described by "candour" "liberal" and "ingenuous"; bad ones by "coxcomb" and "artificial" ([27], p. 39).

Thirty years later, James Parkinson echoed many of Gregory's concerns in a manual intended for the guidance of hospital pupils. He too faulted manners as a source of deception. A medical man, aided by those manners which intercourse with the busy world creates, will draw a favourable attention.

However, these nice manners "induce the multitude to believe him to be fully in possession of that professional knowledge in which he is so miserably deficient" ([41], p. 23). Conventional good manners thus concealed a lack of professional skill.

As in his instructions to his daughters, Gregory linked "servility of manners" and "abject flattery" to a critique of luxury, faulting those who were taken in by rank and fortune:

The external magnificence and splendor which surround high rank, seems to dazzle the understandings of those who live at a distance from it, and who are ignorant what a contemptible inside it often covers ([27], p. 52).

For Gregory, moral worth was almost incompatible with luxury, which was equated with dissipation.

However, Gregory did not discard all manners as artifice. While he chastised some physicians for their indulgence in "a certain nicety and refined delicacy," nevertheless,

As to the general character of a physician's manners, I see no reason why they should be different from those of a gentleman.

Gregory is well aware that manners are socially constructed, and points to the indeterminacy of concepts such as decorum, propriety, and decency. Because such ideas are rooted both in "nature and common sense" and in "caprice, fashion and the customs of particular nations," Gregory adopts an almost instrumental sense of manners ([27], pp. 30–31). While those precepts deriving from nature/common sense are absolute, those which are contingent are less binding.

General norms of acceptable behavior govern many medical situations. The physician who seduces his female patient, for example,

... is a mean and unworthy betrayer of his charge, or of that weakness which it was his duty, as a man of honour, to conceal and protect ([27], p. 28).

Gregory's argument here would have been similar to that employed against Archibald Cleland thirty years earlier: What made Cleland's acts unacceptable was a general code of gentlemanly behavior towards women, not a medical ethics about touching patients. However, unlike earlier writers, Gregory also claimed that medical men had particular standards of behavior due to their profession,

I proceed now to make some observations on the peculiar decorums and attentions suitable to a physician ... ([27], p. 30).

These codes of behavior specific to medicine were "moral duties" and encompassed qualities like candour and temperance ([27], p. 29). In a similar fashion, James Parkinson, in *The Hospital Pupil* spent pages detailing the personal qualities specifically needed for a medical career, such as "A sympathetic concern, and a tender interest for the sufferings of others" and a lack of levity and self-love ([41], pp. 11–17).

In sum, Gregory's prescriptions on manners seem to derive from concerns of Scottish Enlightenment thinkers. Luxury and associated insincerity, which were manifested in excessively artificial manners, were counterpoised to virtue. True moral sentiments would make themselves apparent in those of merit. While manners were still significant, general guides to gentlemanly conduct were no longer sufficient for medical men. Instead, their conduct had to be rooted in an ethic peculiar to their profession.

A much less well-known figure than Gregory made the connections between civic virtue and sincerity even clearer. Peter Reid, an Edinburgh physician, wrote a small guide for medical students, in the form of a letter published in 1809 [47]. It is largely a diatribe against the insincerity of foppish-mannered physicians. Reid inveighs against

[T]his creature of grimace, in straining after that elegance of manners, which is so engaging when it is the expression of real delicacy of sentiment, for want of this necessary fineness of perception, works itself into the most grotesque motions imaginable, and becomes a most valuable specimen of the ridiculous ... ([47], p. 50).

Again and again, Reid denigrates the coxcombs whose public and private behaviors are so divergent.

When Reid discusses medical professors, he again links virtue and sincerity. If professors have any frailties, they

... are more allied to the blunt simplicity of an independent scholar, than the sneaking arts of a worldling; spring rather from the honest arts of nature, than from the nauseous affectations of an artificial character ([47], p. 29).

Hence, the intellectual worth of a professor is not immediately perceptible from his manners. Reid has taken Gregory's precepts one step further and suggested that the duplicity inherent in good manners might imply that true merit was concealed. Despite the peculiarities of the intellectually meritorious, professors are worthy of the students' respect. These men have been nominated by the legislature to the task of "watching over the very source of national honour" ([47], p. 28). Thus Reid links sincerity, intellectual worth, and the foundations of a distinctive Scottish national identity based upon virtue.

IV. CHARIOTS OF HIRE: COMMERCIALIZATION AND MEDICINE

There is another source to the decline of manners, far more concrete than these intellectual and literary worlds. While many medical men trained in Scotland in the latter half of the century, and imbibed some form of the Enlightenment critique firsthand, developments closer to home were also significant. Manners had been transformed into a commodity. Everyone knew that manners could be bought and sold, and were thus an uncertain indicator of gentle status or civic virtue. Anyone reading the little books published by the rash of dancing masters and writing masters realized that civility was only an outward accoutrement, not a moral quality. While some of the general comments about the sale of manners could have been written at almost any point in the past two centuries, the extent to which commercialization had created a huge middle-class market for civility was new.

Even William Chamberlaine, author of the manners-based conduct manual for apprentices, understood manners as a commodity. He advised:

> There is a way of talking off [sic] patients and customers not difficult to be learned by one who gives his mind to it; it is the art of pleasing without seeming to lay ones-self out for it: it very much benefits the Master ([11], p. 171).

Once the apprentice became a master in turn, these skills would be useful to him:

> ... he will have been so far habituated to a pleasing manner of address, that he will have as it were a ready-made stock of politeness to begin business with for himself ([11], p. 171).

Thus Chamberlaine illustrates some of the contradictions inherent in a manners-based ethic. While he selected his own apprentices on the basis of manners, equating morals with manners – saucy answers meant insubordination – he was not unaware that manners could be bought and sold. Indeed, as his telling metaphor suggests, manners could be stored up and dispensed, like any other item in the shop.

So too, the skills of carrying one's body correctly, or a gentlemanly deportment, were for sale. The dancing master Stephen Philpot advised:

... it is the graceful Motion of the Body in walking, reaching out the Hand, Bowing, or performing the other common Actions of Life, in a free, easy, and genteel Manner, that distinguishes the well bred Person from a Clown.

And, of course, Philpot was eager to instruct youth in these gentle attributes. Dancing masters and writing masters were in great demand, both by schools and individual middle-class families, because they could teach these valuable social graces ([21], p. 58). And, of course, it was such families that ambitious surgeon-apothecaries wanted to acquire as patients. No longer were demeanor and address, in the words of John Locke, "the language whereby that internal civility of the mind is expressed" ([34], p. 111). Quite the contrary: Chesterfield pointed out that a man with a moderate degree of knowledge but a graceful bearing would go further in the world that one with "sound sense and profound knowledge" but little grace ([8], p. 150).

The decline of manners is apparent within the medical world through various satires and ironic comments which make a mockery of earlier advice to practitioners, showing how the shared codes of behavior which governed medical men at mid-century were bankrupt by the turn of the century. And, as we will see, some of the ways in which medical manners were mocked point to the process of commercialization which had undermined the whole discourse.

For example, by the 1770s, doctors' attentions to their own appearances were becoming a standard way of poking fun at medical pretensions, and guidebooks came to adopt a more cautious line than those earlier ones which could advise on wigs with a straight face. James MaKittrick, for instance, suggested in 1772 that some peevish individuals claimed that success in the medical profession was earned "by sauntering in coffee houses, or tippling in clubs; – by the size of their wig, by jauntiness of air, prettiness of manner" ([38], p. xxxix). In 1813, the pseudonymous Peter MacFloggem published his parody of medical education guides, entitled *Aesculapian Secrets Revealed*, in which he advised the new apothecary:

your *hair* should be cut in the most elegant stile, *a la physicien*, with a neat queue dangling on your shoulders, or a *six tier* curled peruke; and the whole finely cauliflowered.

Indeed, MacFloggem attacked the basis of manners entirely, ironically advising:

... combine with these qualifications – an obsequious, *cringing*, *fawning*, manner,

> ready with a *bow* and a *smirking* grimace to every miserable wretch from whom you can *wring a fee* ([37], p. 22).

If his advice were followed, and the young medic were to acquire a good reputation, his success would be assured.

John Gregory also faulted medical men for excessive attention to personal appearance. He argued that

> Among the peculiar decorums of a physician's character, much regard has been had to a certain formality in dress, and a particular gravity and stateliness in the general course of behavior ([27], p. 47).

But Gregory employs his distinction between manners derived from nature/common sense and those from fashion to suggest that medical attire falls into the second category. Judges and magistrates should dress with pomp and formality because such clothes promote respect for the law. But in medicine, "personal merit" is what matters and this bears no relation to clothes. Indeed, external formalities "frequently supplant real worth and genius" ([27], p. 49). However, after this diatribe, Gregory grudgingly admits that should all the medical men of one locality dress in a particular manner, it would be unfortunate to deviate from that norm.

Medical men's pretensions to gentility in travelling by coach were also made fun of by the early nineteenth century. In a sarcastic article in the *Edinburgh Medical and Surgical Journal* of 1810, young practitioners were advised to

> Hire a chariot, and put a smart livery upon a bill-stick, to ride behind you ... for nobody in their senses, in London, will send for a walking physician [6].

Of course, mere irony does not imply that a custom has passed away, and medical men continued to pay close attention to their appearances on the road. But the sarcasm of Peter MacFloggem makes it clearer why advice about carriages, once serious, was now a matter for jokes. He starts by describing a carriage as "the infallible passport to the most elevated ranks in society" and then goes on to tell practitioners how to obtain said passport on the cheap. After all, there are "many aspiring *blades*, who would not hesitate to climb from the *counter to the chariot*" if they could afford such emblems of gentle status. But one could hire a man for little and kit him out in secondhand livery at Monmouth Street, along with one's shop drudge, who could be mounted as a footman on the back of a cheap hired-by-the-week coach ([37], pp. 218–220). No longer could a James Nelson be horrified by the deception implied by an apprentice wearing his coachman father's lace ornament. No longer could appearances be considered an "honest bribe" in those words of advice from 1715 – no longer was there any such oxymoron as an honest bribe.

While the commercialization of good manners contributed to the downfall of an ethic of manners, so too did the changing contexts of the practice of medicine. Where James Nelson and William Chamberlaine were obviously writing for a shop-based trade, Peter Reid or John Gregory or James Parkinson addressed medical men who considered themselves members of a profession. In part, such a shift was due to the new forms of practice in hospitals or as poor-law contract surgeons. But the creation of a professional self-identity, a process as yet poorly understood, was not merely the function of hospital practice [10], [23], [33], [35].

Peter Reid, for example, clearly addressed himself to fellow-members of a profession. At the beginning of his essay, he wrote, "medicine has unquestionable claims to the dignity of a liberal profession" ([47], p. 5). Such assertions, of course, prompt the reader to doubt their veracity – if such claims were unquestionable, they would not need mentioning. Instead, it is instructive to see Reid's text creating a form of professional identity. He details disgraces to the profession, and indulges in boundary maintenance, discrediting forms of quackery.

Reid turned a slur against the concept of a profession into a positive quality in discussing medical students' unfortunate predilection for "polite literature." Students read it because it

> ... supplies them with that cant of criticism which enables so many to talk about what they never felt, gives them the forms of literature without the soul, and fits them to hold a part in the literary prattle at a tea table; while it guards against the silly whisper, *He is a mere professional man* ([47], p. 18).

Reid thus combines his usual critique of insincerity with a defense of the concept of a professional man – for him, virtue is rooted in profession.

Similarly, in the writings of Parkinson and Gregory, the authors create themselves anew in the construct of the profession. For them, "profession" lacked modern sociological overtones, instead being closer perhaps to "vocation." What concerned them was, as in the case of Reid, how one could discern true merit (hence their discussions of manners) but also what a profession meant. Issues such as the role of self-interest and the functions of a corporation were never far from sight. In some sense, these men were self-fashioning, moving from a Jewsonian client/patron form of medical practice towards a new type of doctor/patient interaction [26]. What is significant here is the ways in which concepts of profession militated against behavioral norms constructed around general concepts of manners and promoted peculiarly medical codes of conduct.

Where William Chamberlaine schooled his apprentice in those niceties of behavior which would win him patrons, Parkinson, Gregory, and Reid inveigh against such tricks. Parkinson, for example, cautioned:

> It is not for you to cringe for employment, or to lick the ground on obtaining it; for no great obligation of gratitude is due you from any one who chuses to employ you ... ([41], pp. 102–103).

This passage of wishful thinking told doctors to remember that they were in charge:

> ... a physician should support a proper dignity and authority with his patients, for their sakes as well as his own ([27], p. 22).

Rather than seeking to be hired "because it is fashionable," Gregory pleaded that doctors should ground their behavior in their own sense of the dignity of the profession ([27], p. 25). As Reid added, success and honour depended upon skill, since, "in our profession, merit has no direct way of appealing to popular feeling" ([47], p. 41).

Two connected problems particularly troubled these moralists in their attempts to re-make themselves in a professional mould. The first was the question of self-interest – to what extent should medical men act for the benefit of the profession rather than for themselves? Second was the relationship of medicine to trade, since apothecaries and surgeons who ran shops were obviously often motivated by self-interest. Gregory noted that medicine was "either an art the most beneficial and important to mankind, or a trade by which a considerable body of men gain their subsistence" ([27], p. 9). The reader quickly realizes which definition Gregory adopted. Like Reid, Gregory located the dignity of the profession in the learning and knowledge of its individual members, far from the marketplace. Thus he argued,

> ... dignity is not to be supported by a narrow, selfish, corporation spirit, by a peculiar formality in dress and manners, or by affected airs of mystery and self-importance ([27], p. 182).

Instead, "superior learning," "liberal manners," and "openness and candour" by those who profess medicine create the dignity of the profession and distance it from trade.

Such protestations were, of course, motivated by self-interest. As other historians of ethics have noted, the creation of various codes has been associated with shifts in power within the profession. Here I have tried to suggest that the concept of an ethics of medicine is rooted in the late-eighteenth-century creation of a profession in the minds of physicians such as such Reid and Gregory. But this is not a functionalist argument which claims that professional needs predicated the development of ethics. Instead, the creation of a particular conception of a profession predicated an ethic, one derived in part from critiques developed in the Scottish Enlightenment, but also constructed in opposition to the codes of conduct which governed the trade of medicine.

I sum up by quoting the words of an anonymous author of an essay in 1715, which ring true for much of the century. He advised the medical man, "your Employment is Gentleman-like or Mechanical, according as you use people" ([2], p. 97). For most of the century the way in which practitioners heeded this advice was to ground their behavior in the larger and venerable discourse on manners. Archibald Cleland, who had gained his post in Bath through the patronage of the Earl of Chesterfield, lost his position because he understood the message of his patron but poorly. Writers on manners and courtesy provided precepts for conduct, but more importantly, included the reader within a world of virtue and morality in which the appearance of civility was its reality.

In the post-Chesterfield generation, appearance and reality were uncoupled and courtesy stripped of moral connotation. Manners could be bought and sold, and so could not function as an indicator of virtue. Instead, an ethic peculiar to medicine was created, deriving from critiques made by Scottish Enlightenment thinkers. Yet the creation of a medical ethics was not solely the product of the world of ideas. Changing professional structures which made William Chamberlaine's shop-based medicine increasingly obsolete also destroyed the codes of conduct by which he practiced. Nor did his careful instruction in the art of pleasing the consumer have relevance to those who sought to ground their occupation in the institution of the profession.

ACKNOWLEDGEMENTS

I gratefully acknowledge the financial support of the Wellcome Trust whilst I worked on this project. Thanks also to the editors and to Stephen

Jacyna for advice on earlier drafts of this essay.

Institute of the History of Medicine
The Johns Hopkins University

REFERENCES

1. Amussen, S.: 1988, *An Ordered Society: Gender and Class in Early Modern England*, Basil Blackwell, Oxford.
2. [Anon.]: 1715, *An Essay Upon the Duty of Physicians, and Patients, and the Prudentials of Practice*, GJ for Henry Clements, London, p. 338.
3. [Anon.]: 1744, *A Letter to Mr. Archibald Cleland*, J. Roberts, London.
4. [Anon.]: 1744, *A Short Vindication of the Proceedings of the Governors ...*, J. Leake, Bath.
5. [Anon.]: 1775, *An Essay on Politeness* ... by a Young Gentleman, B. Law, London, p. 338.
6. [Anon.]: 1809, 'Hints to Young Practitioners', *Edinburgh Medical and Surgical Journal* 5.
7. [Anon.] [Jones, E.]: 1737, *The Man of Manners: or Plebian Polish'd. Being Plain and Familiar Rules for a modest and genteel behavior, on most of the ordinary occasions of life*, printed for J. Roberts and sold by E. Withers, London.
8. Brauer, G.: 1959, *The Education of a Gentleman, Theories of Gentlemanly Education in England 1660–1775*, Bookman Associates, New York.
9. Burnby, J.: 1983, *A Study of the English Apothecary from 1660–1760, Medical History* Supplement 3, London.
10. Bynum, W.F.: 1985, 'Physicians, Hospitals and Career Structures in Eighteenth Century London' in W. F. Bynum and Roy Porter (eds.), *William Hunter and the Medical World of the Eighteenth Century*, Cambridge University Press, Cambridge.
11. Chamberlaine, W.: 1812, *Tirocinium Medicum; or a Dissertation on the Duties of Youth Apprenticed to the Medical Profession*, printed for the author, London.
12. Childs, F.: 1984, *Prescriptions for Manners in English Courtesy Literature 1590–1760 and Their Social Implications*, Oxford, D. Phil.
13. Clark, A.: 1987, *Women's Silence, Men's Violence, Sexual Assault in England 1770–1845*, Pandora, London.
14. Cleland, A.: 1743, *An Appeal to the Publick ...*, A. Dodd, London.
15. Cleland, A.: 1744, *A Full Vindication of Mr. Cleland's Appeal ...*, W. Frederick, Bath.
16. Collyer, J.: 1761, *The Parents and Guardians Directory*, R. Griffiths, London.
17. Curtin, M.: 1985, 'A Question of Manners: Status and Gender in Etiquette and Courtesy', *Journal of Modern History* 57, 395–423.
18. de Mandeville, B. [1723] 1970, *The Fable of the Bees*, (ed.) P. Harth, Penguin, Harmondsworth.
19. Defoe, D.: 1724 *The Great Law of Subordination Consider'd*, London.

20. Dwyer, J.: 1989, 'Clio and Ethics: Practical Morality in Enlightened Scotland', *The Eighteenth Century* 30, 45–72.
21. Earle, P.: 1989, *The Making of the English Middle Class*, Methuen, London.
22. Elias, N.: 1978, *The Civilizing Process: Volume 1, The History of Manners*, trans. E. Jephcott, Oxford University Press, Oxford.
23. Fissell, M.E.: 1991 *Patients, Power and the Poor in Eighteenth-Century Bristol*, Cambridge University Press, Cambridge.
24. Ford, J. (ed.): 1987, *A Medical Student at St Thomas's Hospital, 1801–1802. The Weekes Family Letters*, Medical History Supplement 7.
25. Goffman, E.: 1956, *The Presentation of Self in Everyday Life*, Edinburgh.
26. Greenblatt, S.: 1980, *Renaissance Self-Fashioning from More to Shakespeare*, University of Chicago Press, Chicago.
27. [Gregory, J.]: 1770, *Observations on the Duties and Offices of a Physician*, W. Strahan and T. Cadell, London. [This is the anonymous edition which was assembled from lecture notes. Quotes are from this edition, but have been cross-checked with Gregory's own text of 1772.]
28. Gregory, J.: 1774, *A Father's Legacy to His Daughters*, W. Strahan, London.
29. Harley, D.: 1990, 'Honour and Property: the Structure of Professional Disputes in Eighteenth-Century English Medicine', in A. Cunningham and R. French (eds.), *The Medical Enlightenment of the Eighteenth Century*, Cambridge University Press, Cambridge, pp. 138–164.
30. Klein, L. E.: 1989, 'Liberty, Manners, and Politeness in Early Eighteenth-Century England', *Historical Journal* 32, 583–605.
31. Lane, J.: 1985, 'The Role of Apprenticeship in Eighteenth-Century Medical Education' in W. F. Bynum and R. Porter, (eds.), *William Hunter and Eighteenth-Century Medical World*, Cambridge University Press, Cambridge, pp. 57–103.
32. Lane, J.: 1988, 'Provincial Medical Apprentices and Masters in Early Modern England', *Eighteenth-Century Studies*, 14–27.
33. Lawrence, S. C.: 1985, *Science and Medicine at the London Hospitals: the Development of Teaching and Research, 1750–1815*, PhD thesis, University of Toronto.
34. Locke, J.: 1693, *Some Thoughts Concerning Education*, in *The Educational Writings of John Locke* (ed.), John William Adamson, Cambridge, 1922.
35. Loudon, I.: 1986, *Medical Care and the General Practitioner, 1750–1850*, Oxford University Press, Oxford.
36. Lucas, J.: 1800, *A Candid Inquiry into the Education, Qualifications, and Office of a Surgeon-Apothecary*, S. Hazard, Bath.
37. MacFlogg'em, P.: n.d., *Aesculapian Secrets Revealed*, C. Chapple, London.
38. MaKittrick, J.: 1772, *Commentaries on the Principles and Practice of Physic ... To which is Prefixed, An Essay on the Education, and Duties of Medical Men*, T. Becket, London.
39. Mason, J.: 1935, *Gentlefolk in the Making, Studies in the History of English Courtesy Literature and Related Topics from 1531 to 1774*, University of Pennsylvania Press, Philadelphia.
40. Nelson, J.: 1753, *An Essay on the Government of Children under Three General*

Heads: viz. Health Manners and Education, R. and J. Dodsley, London.
41. Parkinson, J.: 1800, *The Hospital Pupil*, H. D. Symonds, London.
42. Phillipson, N. T.: 1973, 'Towards a Definition of the Scottish Enlightenment', in P. Fritz and D. Williams (eds.), *City and Society in the Eighteenth Century*, Hakkest, Toronto, pp. 125–147.
43. Philpot, S.: 1747, *An Essay on the Advantage of a Polite Education Joined with a Learned One*, printed for the author, London.
44. Pocock, J. G. A.: 1975, *The Machiavellian Moment: Florentine Political Thought and the Atlantic Republican Tradition*, Princeton University Press, Princeton.
45. Porter, R.: 1987, 'A Touch of Danger: The Man-Midwife as Sexual Predator', in G.S. Rousseau and R. Porter (eds.), *Sexual Underworlds of the Enlightenment*, Manchester University Press, Manchester, pp. 206–231.
46. Pullen, C.: 1968, 'Lord Chesterfield and Eighteenth-Century Appearance and Reality', *Studies in English Literature, 1500–1800* 8, 501–515.
47. Reid, P.: 1809, *A Letter on the Study of Medicine Addressed to A Student*, Adam Black, Edinburgh.
48. Rolls, R.: 'Archibald Cleland: c. 1700–1771', *British Medical Journal* 288, 1132–1134.
49. Rothblatt, S.: 1976, *Tradition and Change in English Liberal Education*, Faber and Faber, London.
50. Sher, R.: 1985, *Church and University in Enlightenment Scotland: The Moderate Literati of Edinburgh*, Princeton University Press, Princeton.
51. Simon, J.: 1979, 'Childhood in Earlier Seventeenth Century England', in K. S. Dent (ed.), *Informal Agencies of Education*, Proceedings of the 1977 Annual Conference of the History of Education Society of Great Britain, History of Education Society, pp. 1–27.
52. Smollett, T.: 1752, *An Essay on the External Use of Water ...*, M. Cooper, London.
53. Young, S.: 1892, *The Annals of the Barber-Surgeons of London ...*, Blades, East and Blades, London.

CHAPTER 2

DAVID HARLEY

ETHICS AND DISPUTE BEHAVIOR IN THE CAREER OF HENRY BRACKEN OF LANCASTER: SURGEON, PHYSICIAN, AND MANMIDWIFE

When looking at the ways in which societies change, one is drawn to disputes rather in the way that the geologist is drawn to fault lines. Disputes lay bare the points of conflict between and within the strata of a hierarchical society, indicating tensions resulting from shifts in the structure. This is especially important when one wishes to explore the norms and values current within a society, since morality is not generally the subject of much discussion in everyday life, except when conflicts occur.

Historians interested in disputes have recently tended to adopt an outlook inherited from functionalist legal anthropology [1], [7]. This approach focuses on the settlement procedures offered by society in an attempt to preserve the *status quo* rather than on the content of the disputes themselves. It produces an unduly rigid model of society, lacking the dynamic element integral to disputes. If instead the development of social norms over time is the subject of the historian's scrutiny, the ostensible content of the disputes and their social context become as important as any attempts to provide methods of resolving them.

This is particularly true of "medical ethics" in eighteenth-century England since disputes bearing ethical standards were widespread long before professional codes, or quasi-legal settlement mechanisms were established and indeed precipitated the creation of such codes [12]. The principal matter of concern in these disputes was the behavior of the practitioner, especially towards his colleagues. These concerns feature so prominently in Thomas Percival's pioneering work that it has puzzled modern readers, who have gone so far as to argue that what he calls 'medical ethics' falls more properly under the rubric of medical etiquette.

If historians are to understand the ethical position of early modern medical practitioners, it is necessary to consider the ideals to which they aspired, the difficulties that made them fall short, and the means they employed to castigate one another's shortcomings. Even after the introduction of codified systems of medical ethics, the ideal types of

R. Baker, Dorothy Porter and Roy Porter (eds), The Codification of Medical Morality, 47–71.

practitioner tacitly taught to pupils continue to play the greatest part in shaping the ethical choices of medical personnel. Prior to Percival and his successors, ideal types and their social or religious rationales were the only sources of guidance available. In this study, the problems of a single practitioner, Henry Bracken of Lancaster, are examined in detail in order to indicate the kinds of stresses to which a provincial doctor might be exposed in the mid-eighteenth century. This should illuminate the background to Percival's essay and also provide a salutary warning against citing published medical works without considering the strategies and context of the author.

Henry Bracken is best known as the author of a book on midwifery, *The Midwife's Companion*, published in 1737. There is a copy of this rare book in the Wellcome Library; consequently, Bracken's opinions have been cited by historians as different as Thomas Forbes, Angus McLaren, and Valerie Fildes ([11] p. 353; [19] pp. 19–20, 111, 143; [9] pp. 190, 227; [10] pp. 111–113). His life, although far from obscure, is less well known to historians of medicine, so a brief sketch of his education and career will be necessary to provide a context for discussion of his disputes and their ethical implications ([12] pp. 161–164; [6]; [8]).

Born in 1697, the son of a Lancaster innkeeper, Bracken received a grammar school education and was then apprenticed to Thomas Worthington, a member of the prominent Wigan family of Catholic surgeons and physicians that was frequently accused of Jacobite plotting. After instruction in medicine, surgery and chemistry, Bracken went to St. Thomas's Hospital for three months, walking the wards with Drs. Wadsworth and Plumtree. Disappointed at the lack of cadavers, he then went to the Hôtel-Dieu in Paris, where he found plenty of opportunity for dissection and clinical observation. He was enabled to attend the midwifery cases through the good offices of the English ambassador, the Earl of Stair. After Paris, Bracken attended Boerhaave's lectures in Leyden for eighteen months and assisted the great man in his private practice (according to his own account, but although he matriculated at Leyden University on 29 August, 1730, there is no record of his taking a degree). After trying to establish himself in London, Bracken set up practice in Lancaster as physician, surgeon, and manmidwife, remaining there until his death in 1764.

Bracken was married to the daughter of a Lancaster bookseller and the marriage appears to have been a happy one (his wife survived him until 1785). His only son, a promising physician, died of gaol fever contracted while visiting him in Lancaster Castle during his brief imprisonment after the 1745 rebellion. His three daughters also died young. Apart from medicine, Bracken's main interest was horse-breeding and he was one of the best known authors on farriery in England. The only appearance he makes in William Stout's autobiography is when that worthy was knocked down by a runaway horse being exercised by Bracken's stable lad ([18] pp. 236–237). He also bred game-cocks for fighting and he was fond of angling, shooting, and coursing. He made his own malt and brewed beer supplying the Castle; at one time, he supplied the town with coal from Burton in Lonsdale. He also traded in wine and fruit with Portugal and the Levant, in his own ship. Bracken was a staunch Church Tory and became markedly pious in his final illness, although very depressed at the sound of the passing-bell according to his last and most famous pupil, William Moss of Liverpool ([20] pp. 123–124).

Although Bracken was known for his good fellowship and lively wit (especially at the pub in China Lane), he was quick to anger and deeply resented injuries. His disputes were numerous and varied, some leaving extensive documentation and others, through the vicissitudes of archive survival, hardly a trace. His medical quarrels indicate some of the most significant areas of conflict concerning professional conduct within eighteenth-century medicine. Since his medical disputes often had political undercurrents, his legal and political conflicts inevitably affected his standing as a medical practitioner and it would be artificial to separate them into rigid categories.

Bracken is best known for his published works and it is probably most convenient to use them as a starting-point. His three major works (on midwifery, farriery, and the stone) were all published in the late 1730s, probably because he needed publicity if he was to leave Lancaster and set up in London. In his own time, Bracken's best known work was on the care of horses, his various publications on the subject going through many editions, starting with his scathing annotations on Burdon's farriery notes [3]. His farriery writings were famous chiefly because his only serious rival, William Gibson, was a former army surgeon who had none of Bracken's intellectual pretensions.

Before writing his most substantial farriery text, Bracken published his thoughts on midwifery in *The Midwife's Companion,* dating the fulsome dedication to "my Worthy and very Learned Preceptor, Harman Boerhaave," 20 November, 1736. The book is a mine of information about provincial obstetrics, gynecology and paediatrics, but only a few ethical aspects will be considered here. When Bracken wrote, man-midwifery was a relatively recent innovation and the attempts of surgeons to become accepted as the practitioners of first resort among affluent families, instead of merely dealing with emergencies, generated a wide range of ethical issues.

Bracken was no shrinking violet and modest humility is rarely found in his writings, although he acknowledges it as a virtue. After attacking the Chamberlain family's use of secret techniques for delivering children without the use of instruments, "merely to gain Esteem," he boasts that he has "a more sure and safe Manner, than has been taught by Aristotle, Culpeper, Ettmuller, Daventer, Dionis, Maubray, or any other Author yet extant." This he attributes to his training under the excellent Worthington and his further education in Paris. He states that he would have preferred to write in the learned style "proper for my Fellow-labourers the Men-Midwives" but has chosen instead to write in a more universally comprehensible manner ([5] sig.A4). Undoubtedly the reason he chose to write for the general public was that he was concerned to attack rivals and promote his own practice. Writing for an audience of colleagues was only an effective strategy when one lived surrounded by them and could benefit from their referral of patients.

Bracken's criticism of self-sufficient midwives is well-known. His self-recommendation was couched in indignation against the incompetence of female competitors. He attacks "bold, daring, presumptuous, though ignorant Midwives, who pretend they have no Occasion for the Assistance of a man; but will themselves venture to use Instruments to extract the Child." He asserts that such "Butcherly Midwives" would happily deliver a child "either Whole or Piece-meal" rather than seek assistance, and he calls for "judicial enquiry into the Circumstances of the Case of Children born dead, maimed, or distorted: But so far from this, that the Law is such at present, That a Woman who can only procure the Hands of a few good natur'd Ladies, or Justices of the Peace, to recommend her to the Bishop or Ordinary, shall have a Licence to Practise, although neither those who recommended, nor the Bishop himself know anything of the Matter" ([5] sig.A5–6).

One of the issues raised here was licensing. The extant ecclesiastical system of licensing required proof of competence, but it was in decay by this period. It had always mainly been concerned with the moral character of applicants and the maintenance of ethical standards rather than with strict enquiry into midwives' technical knowledge. Bracken's views were not calculated to increase his popularity among either the midwives or the ecclesiastical lawyers of Lancaster and so when he complains that, in sixteen years of practice in the area, he had only been called four or five times without the prior presence of a midwife, we should, perhaps, take the complaint with a grain of salt. He blames the lack of early consultation on the excessive modesty of women. Perhaps a more ingratiating manner would have encouraged earlier calls, but the establishment of a clientele using prior booking was never easy. Bracken is also quick to condemn a rural midwife as "but an ignorant Woman" for omitting to send a clinical description when she required assistance but he fails to make any suggestions as to how the technical education of midwives or their co-operation with surgeons might be improved ([5] pp. 64–65, 122; [24]).

Less well known than his criticism of midwives is his attitude towards other male practitioners, especially those whose knowledge was more theoretical than practical. "I cannot imagine why any Woman should apply to a Physician (if he be meerly such) to know if she's with Child; for it is quite out of their Way of Business, unless they practise Midwifery; because they are not used to the handling of Women's Bellies" ([5] p. 36). He is highly critical of the increasingly fashionable technique of podalic version, which he sees as designed for the convenience and glory of practitioners, regardless of the danger to the child. He tells the tale of a recent case at Preston when the midwife called a young practitioner who, "having read something about extracting Children by the Feet," pulled out the living child and left the head behind. Bracken compares the man with his least favorite kind of competitor, "a young Physician, just come from College, with his Head brim-full of the Theory, prescribing hab-nab at random, thinking there is nothing more to do than, like a Juggler, to pronounce Hey Presto, be gone, and the Distemper is to vanish in the twinkling of an eye" ([5] pp. 65, 163, 171–172). Throughout his works, Bracken rails against the arrogance of practitioners who rely upon book-learning rather than experience, neglecting the welfare of their patients in order to impress their audience.

The *character* of the practitioner is of prime importance in the promotion of ethical conduct towards both patients and colleagues. Bracken is clear that the good character of a man-midwife is central to the success of his practice and he lists the necessary attributes, such as sobriety of dress and manner, physical strength and dexterity, and so on. He is especially vexed by the reputation of his colleagues for butchery.

Moral qualities were, of course, especially crucial for practitioners of manmidwifery, if their novel practice was to be accepted by their neighbors for routine births rather than just in emergencies. Bracken himself was not renowned for sobriety of manner and it is notable, in view of contemporary anxieties about the sexual threat posed by man-midwives, that even his biographer admitted that Bracken was "addicted to unlawful commerce with the sex" ([5] pp. 135–141; [6] p. 102; [22]). (Perhaps the ladies of Lancashire had good reason to be concerned for their modesty. Bracken was named as the guilty man in a 1734 Fornication case.)

The worst characteristics of both male and female practitioners, in Bracken's view, were undoubtedly haste, arrogance, and an undue eagerness to use instruments, which should only be used when the head had become detached from the body. "As for my own part I can sincerely speak it, and of which my Neighbors can testify the Truth, that I bring all Children into the World whole and entire, without tearing them in pieces as is the common Practice in difficult Births" ([5] pp. 149–154, 165). Bracken's relations with urban midwives were evidently not as bad as with their rural sisters. It would have been necessary for him to work alongside them if he was to have any midwifery practice at all, and he appears to have had a reasonable working relationship with a Mrs. Haresnape.

In addition to criticizing physicians and midwives, Bracken expatiates in his writings on the unreliability of juries of matrons, called upon to assess whether condemned women were pregnant. He then recounts a case of infanticide where the woman was acquitted. "I was call'd to view the dead Child," but his evidence was not called at the assizes. He insists that had he testified the woman would have been hung ([5] pp. 35–36). The case in question occurred when Bracken was one of the bailiffs, in 1730–31, so it is perhaps not surprising that he was called as an expert witness since he was responsible for paying the coroner if a verdict of unlawful death was passed. The borough accounts contain an entry, "Allowed them pd. Mr Mayor's fee as Coroner upon sitting the Inquisicion over Anne Walmisley's Child Supposed to be murdered, 0:13:4." The murder verdict of the coroner's jury was duly sent to the grand jury who found it *billa vera* but, when the case was brought to trial, the petty jury found her not guilty.

Bracken underestimated the difficulty of securing a conviction for this crime. When he himself acted as *ex officio* coroner of Lancaster during his 1748 mayoralty, he heard a case of infanticide. The inquest heard the evidence of two women summoned to the scene by the accused woman's employer, the wife of a Quernmore husbandman and a Lancaster midwife, Sarah, the wife of George Haresnape of Moor Lane. No other expert witness was called. The accused was acquitted by the trial jury, with exactly the same verdict as was passed in every trial for infanticide in Lancashire during the long reign of George II.[1] When writing his book (a decade earlier) he apparently failed to realize that it was not the quality of the medical evidence that principally determined whether or not a woman would be condemned at this time, but the attitude taken by the trial jury. Coroners had a pecuniary interest in a finding of murder and neither their juries nor the grand jury had the responsibility of condemning young women to death. They could all give free rein to their moral indignation. Trial juries were more inclined to make the most of any favorable circumstance, however convincing the evidence [13].

Bracken's next book, *Farriery Improved*, was published a few months after his book on midwifery. The long list of subscribers included a host of medical men, but Bracken still felt obliged to defend the propriety of his applying Newtonian medicine to the treatment of horses: "Therefore, I say, let not my Brethren complain at me, as if I were debasing the Profession seeing it is certainly fact, that he who cannot write sensibly about the Distempers in brute Creatures, is not fitly qualified to prescribe for Man, by reason 'tis plain he has not studied Nature throughly" ([2] pp. ix–x). At a stroke, Bracken turns his text from mere advertisement into a positive recommendation. This defensive strategy suggests that he had been criticized by his competitors for devoting too much of his time to the study of horses.

Within the text of the book, Bracken dismisses both traditional remedies and the authors who retail them, emphasizing the importance of applying the same kind of rationality to veterinary medicine as would be applied to human medicine by regular physicians. That rationality was based on mathematics and the sciences, the Newtonian project in medicine.

> I must ingenuously confess that my Faith is not pin'd upon the Number of Ingredients in any Composition, but rather upon such Things as are particularly adapted to the Cure of the Disease, and for the Administration of which there is a Why and a Wherefore; for really there is no end of the Practice which is founded upon Nostrums, seeing every one is bigotted to his own Receipt. Therefore it is well with us that the Art of Medicine is at this time (by the Helps which accrue, from Mathematicks,

Mechanicks, Chymistry, and Natural Philosophy) brought to such a Certainty, that all Family Receipts are thrown out of Doors, and sufficiently exploded as superstitiously and ridiculously contrived ([2] p. 165).

As his strident advocacy of rationality in medicine suggests, Bracken had little good to say about empirics. He had especially harsh words for the eye surgeon Chevalier Taylor, despite his evident ability, because he is boasting of an ability to do more than he could in fact do ([2] pp. 175–178, 182–187, 189, 215–216, 224). Nevertheless, Bracken acknowledges that empirics possess remedies and techniques that are sometimes more useful than those used by regular practitioners. Like Dr. Thomas Beddoes after him, he explains this in terms of the commercialization of medicine. Thus, after discussing the utility of mercury and antimony in skilled hands, Bracken condemns his contemporaries: "And no doubt but at this Time there are in the Hands of some Empiricks, Medicines upon this Basis which are too much a Shame to Learning and Education. But how can it be otherwise, when Practice is calculated rather for the Ease and Dispatch of a great deal of Business than to cure Patients; whereby Cases of Difficulty are not worth the Care of any but Quacks, and such as venture at Random." Other effective techniques, such as cautery, were only used by empirics, despite good Arabic authority. "But, when Physicians and Surgeons are timorous, either for the present Ease of their Patients or in Diffidence to their own Judgments, we must expect several Cures will be perform'd by Empiricks and Quacks, who venture upon rougher, yet more effectual Means." Antimony was excluded from the shops "to oblige the present delicate Age, who rather choose what they ignorantly style a *safe Man*, than one who prescribes more effectual Medicines, and moves upon a more certain Basis, and much better Philosophy" ([2] pp. 395, 543–545).

Clearly, Bracken set his face against two major influences on the practice and ethics of eighteenth-century medicine, the power of the consumer in the marketplace and the development of Whig delicacy and sensibility. Rather, he looked back to a more robust style of medical practice, admiring Thomas Sydenham and John Radcliffe rather than Hans Sloane and Richard Mead. Sydenham's radicalism was being forgotten and he was turning into a Tory hero, fit to be praised by Samuel Johnson, a fate that also befell Boerhaave. Bracken compares Sydenham's candid admission of juvenile error with the behavior of the latest generation of graduate physicians: "But so far from following this glorious Example are many of your present Theoretical Gentlemen, who just dispatch'd from Colleges, brim-full of Diploma and Ignorance, foist their Nonsense and Absurdities upon the Publick, yet never own their Errors, but imprudently and positively insist upon the Justness of their juvenile and shallow Prognosticks in the Cure of Diseases, insomuch that an old experienced Practitioner has but a poor Chance of being heard, unless the Audience happen to be better Judges than are commonly met with on such Occasions" ([2] pp. 459–460).

Bracken might have been sensitive to the slights inflicted on him by callow upstarts but he met his match in Peter Kennedy, whose work on eye diseases he cites with some respect in *Farriery Improved* ([2] pp. 183–185). In the January following publication, Kennedy wrote a letter demanding alterations to the text. Bracken's reply was placatory but unyielding on the technical point in question so the correspondence became increasingly acrimonious. Bracken suggested arbitration to resolve the dispute. "I shall come to Town yet, and I think to settle at London; therefore I must defer my further answer to your Queries 'till then...." Kennedy became vitriolic, boasting of unfamiliarity with Bracken's works. Bracken accused Kennedy of being old and crabbed, a plagiarist and insignificant. Kennedy was cut to the quick and produced a long self-defence, whining about his lack of influential patrons and accusing Bracken of stealing ideas from him. He also wrote sarcastic "Rules for the Use of Dr Bracken, when he comes to Town," advising him to get custom and fame by associating with apothecaries and midwives. The whole correspondence was published by Kennedy in 1739, the year of Bracken's next book [15].

If Kennedy's attack and his own failure to stay calm had not destroyed all possibility of him successfully moving to London, the response to his next book almost certainly did. Bracken's *Lithiasis Anglicana* was an overly bold and soon forgotten venture into a partisan debate. A group of London scientists, spurred on by David Hartley and Stephen Hales, succeeded in persuading Parliament to pay handsomely for the recipe of Mrs. Joanna Stephens's cure for the stone. It was to be assessed through *in vitro* experiments and testimonials from grateful patients. The project was mainly supported by Whig physicians and opposed by Tory surgeons. The only practical effects of years of international research were to be the establishment of David Hartley's career, the creation of large collections of concretions, and the development of improved laboratory techniques [23].

Provincial physicians such as Bracken in Lancaster and Brownrigg in

Whitehaven were obliged to take notice of the celebrated lithontriptic as a result of pressure exerted by patients unwilling to submit to the surgeon's knife. Among the patients listed in Hartley's book of 1739 on the medicine were several in North West England, including Ambrose Gillison of Lancaster ([14] pp. 72, 102, 104–105, 108–109, 118–120). Few provincial authors were rash enough to enter the lists. Hartley himself was not a medical graduate at this time and had been summoned by the Royal College of Physicians in 1737, escaping only because he was suffering from the attack of gravel that aroused his interest in the medicine.[2] Nevertheless, his project was supported by leading Whigs in Parliament and the Royal Society.

Attacking the pet enthusiasm of such people was a forlorn hope, at least at this early stage. Bracken probably saw Hartley as just another Cambridge-educated novice dabbling in medicine and he denounced the vaunted cure as simply a quack remedy, notable only for its fashionableness and its money-making capacity: "it will scarce be heard of a Year hence; for every Age produces some strange Nostrum or other...." The proposed methods of evaluating it were quite inadequate. Bracken prefers clinical field trials to the laboratory experiments and testimonials that were to be assessed by some parliamentary or merely learned committee. "It would have been better concerted (in my Judgment) if an ingenious Surgeon in every County, together with a Physician or two had some of Mrs. Stephens's Composition sent them to make Trial of upon Patients really afflicted with the Stone in the Bladder...." Experienced provincial practitioners and the London hospitals were the best judges of the potion's efficacy ([4] pp. 2–3). Bracken clearly felt that a vast sum of money had been paid for a useless remedy that was to be assessed by incompetent partisans.

Yet Bracken did not reject the quest for a medical cure.

> I am so far from depreciating or ridiculing any Attempt that may be made for a Dissolvent of the Stone, &c. although I have practised as a Lithotomist, above Twenty Years, with as good Success as the rest of our Fraternity; no, I declare I value the good of Mankind, more than to wish them so ill as to hazard so dangerous an Operation, if the Stone could any way with safety be dissolved ([4] p. 29).

Although his practice of lithotomy was generally successful, according to his own account, Bracken felt himself to be under unreasonably close scrutiny. "I durst not venture out of the common Practice, for fear of Censure and Party Malice, the latter of which I have sufficiently felt a few

Years past; so that by the Way, it is not Politick in one of our Business to join with any Party but rather sit neuter" ([4] p. 45). Bracken gives detailed accounts of his methods, together with his sources for them. "I must own it is good to have Precedents to warrants one's Practice, since Prejudice and Party Malice are grown so high, on account of different Opinions in voting for Members to serve in Parliament; for I was threatened with no less than an Indictment for Murder, if a Patient under my Care, about three Years ago, had died after I had cut him for the Stone" ([4] pp. 51–52). Such an indictment would perhaps have succeeded in passing a politically biased grand jury, but it would have been difficult to obtain a conviction. However, even the threat must have been highly alarming.

No sooner had Bracken published his book than he was attacked by one Omnelio Pitcarne, M.D., probably a paid hack, in terms that indicate the political importance of flattening any opposition to the medicine. He suggests that Bracken's motive was probably self-interest: "Whether 'tis his Apprehensions of Decrease in Business (by the Publication of Mrs. Stephens's Medicines) that mov'd Him to it, I can't directly say...." He accuses Bracken of rejecting the potion because it was discovered by a woman and of calling the members of the House of Commons fools. He calls Bracken's suggestion that the medicine would soon be forgotten, "whimsical Predictions and childish Arguments," and he derides Bracken for respecting the opinion of country practitioners rather than the learned judgement of London gentlemen ([21] pp. 5–11). Having failed to engage with Bracken's arguments, he then lists over a hundred alleged cures, mostly taken from Hartley, whose very failures become successes here.

Why did Bracken want to leave the security of his native town for the uncertainties of London medicine? What is the meaning of Bracken's repeated comments about political disputes impinging on his medical practice? It is necessary to untangle the skeins of Bracken's complicated life through the use of local and legal records if one is to uncover the significance of these remarks. In seeking to understand the ethical stance of a medical practitioner, it is important to examine his character, politics, and religious beliefs.

In the period 1726–31, Bracken acted several times as auditor, chamberlain, or bailiff for the Lancaster Corporation, as his father had done before him. No problems appear to have occurred, to judge from the Borough Accounts Book, but Bracken and his fellow bailiff for the year 1730–31, John Borranskill, became embroiled in a Chancery case brought against them by Ambrose Gillison that dragged on until July 1732.[3] This appears to have been something of a watershed period for Bracken, whose life became increasingly tempestuous. Both national and local politics were becoming increasingly polarized as a result of Tory detestation of Walpole's corrupt government. The composition of the Lancaster Common Council was changing and the Whig merchant, Christopher Butterfield, was elected Mayor for 1732. Henceforth, office-holders were increasingly Whigs or even nonconformists and Bracken did not hold office again until his mayoralty in 1748. His political marginalization was coupled with increasing litigiousness, that paralleled the vexed tone of much of his writing. Bracken was involved in twelve cases before the borough court between 1734 and 1747, either as plaintiff or defendant.[4] Several of Bracken's opponents can be identified as Whig lawyers but details of the cases do not survive. They were presumably petty squabbles with an underlying political motivation, like his attempt to get the grand jury to indict John Bryer for obtaining £10 from him under false pretences.[5]

An example of a political adversary who also became a personal enemy in the tense atmosphere of the 1730s was the lawyer Robert Chippindale, a tenant of Bracken who was to become the borough attorney in 1738.[6] Chippindale, his wife, and his lodger were involved in a series of threats against Bracken that resulted in violence, sureties to keep the peace and indictments. According to statements signed in December 1736 by Anne Bracken and her servant, Elizabeth Parker, they had been threatened late on Saturday night by Richard Hoggart of Holme, Westmorland, yeoman. The servant had been pinched and shoved in the street and Mrs. Bracken had been called a "nasty Bitch" and threatened with a fist, after her door had been broken down.

Richard Hoggart, who described himself as a gentleman, Agnes, wife of Robert Chippendale, and her servant, Mary Ball, made a statement on the same day, accusing Dr. Bracken of being guilty of a violent response to this affront. According to them, after the words had passed between Mrs. Bracken and Hoggart, who lodged with the Chippindales, Bracken came into their kitchen and beat Hoggart with a large lead-weighted hunting whip until it broke and Hoggart was lying prostrate and bloody. He slashed at Mrs. Chippendale when she intervened. Hoggart attempted to escape to the Mermaid Inn and was attacked on the way. At the inn, Bracken "presented a pistoll to his breast and struck the Lock." Hoggart, "terrifyed, said Lord Doctor what will you murther me...to which the said

Bracken Replyed Yes God Damn you you Dogg I will kill you." The pistol failed to fire, so Bracken beat him with it and the remains of the whip until help arrived.

Hoggart obtained an order for Bracken to keep the peace but Bracken replied with a similar petition. A few days later, a surgeon called Westropp Berry, presumably Bracken's apprentice, and Elizabeth Parker deposed that they had twice heard Robert Chippindale threaten to "set fire to the said Doctor Bracken's House and blow him and his Family up." Chippindale was bound over but the indictment was dismissed in August by the grand jury, dominated by Bracken's political opponents, although they did find a true bill against Hoggart for breaking the door down and insulting Mrs. Bracken. More significantly, they found a true bill against Bracken for the assault and the attempt to shoot Hoggart. Both Bracken and Hoggart submitted and they were fined one shilling each.[7] Elizabeth Parker was indicted for perjury in her evidence and pursued by a succession of special warrants for some years.[8]

At the same assizes in the spring of 1737, Bracken was indicted on two further charges. One was barretry (i.e., vexatious litigation) an unusual accusation to be heard at assizes and one with severe implications for Bracken's good name. According to the indictment, Bracken "was and now is a Common and Turbulent Slanderer Brawler Fighter and Sower of Discord amongst his Neighbors an Oppressor of his Neighbors and a wrong doer So that he hath stirred up and Procured and still stirs up and Procures divers Suits and Controversies...." The case was delayed because a key Crown witness was in London and Bracken was finally acquitted in March, 1739.

The other case was an accusation of forgery. Bracken had allegedly forged the hand and seal of Richard Backhouse to a letter or warrant of attorney, to enable him to collect sums of money from several people named in it, including a bag left in Ulverstone containing £161 3s.6d. He was found not guilty in March 1738.[9] According to Bracken's biographer, this charge arose from the doctor guiding the hand of a patient who was paralytic and intermittently insane. "This was done with the concurrence of the patient while in his right mind, and in the presence of his sister and other friends. The act, therefore, was not only legal, but kind and worthy of praise" ([6] p. 29). Perhaps instead one should say that Bracken acted incautiously and with doubtful legality.

While Bracken was still preparing to face the assault charge, a series of cases began that would have grave consequences for Bracken's reputation. In January, 1736, William Elletson, an infant, by John Holt his nearest friend, began the first of a long series of Chancery actions against Edmund Lancaster and his wife, among others. The name of Bracken did not appear in these actions, except as one of the original commissioners, but it became clear before long that he was really the prime mover. He is unlikely to have made a secret of his involvement.[10] In 1740, Bracken was accused before King's Bench of publishing a scandalous libel on Edmund Lancaster who was in possession of the disputed estate. In March, Bracken had written a letter to a maltman, Richard Meadows, threatening to raise the rent on his kiln because he was supporting Lancaster. Meadows showed the letter to "Mr.Lancaster being a Neighbour and Acquaintance of his and a person of Reputation and Credit in the Neighbourhood...which gave him the said Lancaster and his Familey very great uneasiness." According to the deponents, "the said Bracken hath been very industrious in making it and the contents thereof known." Robert Chippindale testified to the handwriting being Bracken's. Bracken called Lancaster "a most egregious villain and forsw'n. Rascal as he himself is very sensible, and for y^{r} part Friend Meadows, I think you seem to pipe in with the wickedly-inclined in this matter, so shall e'en turn you (amongst the rest) over to *Satan* to be buffeted." At this point in the dispute, Bracken was still able to present himself as the disinterested friend of young Elletson, however little his political opponents might believe him. This was to change with the then imminent arrival of the boy's mother from America.[11]

In May, 1742, Bracken and a Lancaster skinner deposed before a commissioner of the King's Bench that for the last month William Elletson and John Sill, clerk of the attorney John Bryer, had been distributing a scandalous petition against Bracken, making it "public and notorious in a great many Towns and Villages in Lancashire and Elsewhere." They were doing "their utmost endeavour to vilifie and asperse his this Dept. Henry Bracken's Character." The skinner, James Benson, rather improbably deposed that Bracken was "a person of a peaceable disposition and of good name and Reputation."

The petition, written by Sill, was in the name of Frances Elletson and her son William, formerly of Maryland but now of Lancaster. According to them, the boy's uncle had died intestate, leaving an estate worth over £4000. Shortly afterwards, Bracken had sent for the boy "under a pretence of friendship," promising that out of charity he would secure the boy's inheritance for him. Although relations had also invited him, the boy was swayed "by the Earnest sollicitations and fair promises made on behalf of the said Doctor Bracken, by one Ralph Holt who had formerly been an Apprentice to him." He arrived at Liverpool with Holt in October, 1736, but Bracken prevented his relations from seeing him "and conveyed him secretly away, and kept him for two Years or thereabouts in Private at Bury in the said County."

In 1739, he had been persuaded to become Bracken's apprentice, thus giving Bracken "power over your Petitioner's Person." Bracken then commenced law suits about the estate and tried to get Elletson to sign bonds for a large apprenticeship fee and to convey the estate to him "which he at last effected, by threatening to Turn your Petitioner a Begging." After getting control of the estate, Bracken had turned him out, "and left him and his said Mother in the greatest want and Necessity." They were collecting money to get a passage to join a wealthy relation in Jamaica. Over a hundred small donations are listed at the foot of the petition, including the names of all Bracken's Whig enemies such as the Bryers and the Chippindales, the Vicar Dr. Fenton, the physician Dr. Drinkell, the Recorder and the Commissary.[12]

It is difficult to know how many of these allegations were true but the cases dragged on for decades, first against Bracken and then against the trustees who had control of the estate pending the settlement of the dispute between the two nephews of the deceased. As late as 1760, Bracken was still attempting to bring a counter-action against Ralph Holt and William Elletson.[13] Bracken's biographer asserted that the boy, his brother, his mother, and a witness all came over from America together and were supported by the doctor, who bore not only the costs of the boy's education but also the entire costs of the case. The boy was very grateful, we are told, when Bracken secured the estate for him and only failed to repay his benefactor because he made an unfortunate marriage. The doctor's only reward was an accusation of barretry inspired by the dead man's sisters which, "as it deserved, met with nothing but the derision of the Court" ([6] p. 29).

The timing of the two cases makes it unlikely that the Chancery case was the sole cause of the barretry indictment and it is clearly not true that Elletson was grateful for Bracken's charitable endeavors. The truth of the various charges levelled against Bracken in the period preceding the Jacobite rebellion of 1745 is probably impossible to establish at this distance, given the laconic nature of the scattered records generated by English law, but his reputation was evidently not as unblemished as his biographer would like readers to believe.

The events of 1745–46 mark another watershed in Bracken's life. He was imprisoned, his son died, and he seems thereafter to have become a more peaceable man or, as his biographer puts it, "he perhaps never after thoroughly regained his former vivacity" ([6] p. 101). A colorful version of the involvement of Bracken with the Jacobite army, published by his biographer, has generally been accepted by later writers. In brief, the story is that Bracken drank with those of the Pretender's officers who recognized him from his Paris days but preserved his loyalty by an ambiguous toast. Bracken sent word to General Wade of the rebels' numbers and pursued the stragglers with a group of horsemen, capturing a messenger. On the return of the retreating rebels, Bracken's house was looted but his enemies conspired to have him imprisoned as a Jacobite. He was eventually bailed and no evidence was ever produced against him ([6] pp. 29–30, 100–101). The tale is told with a wealth of circumstantial detail, omitted here, and it is not intrinsically incredible. Unfortunately, if no evidence was produced against him, none is likely to remain in the archives.

An examination of Bracken, dated 14 April, 1746, while he was imprisoned in Lancaster Castle, is extant in the records of King's Bench. Possibly taken by Dr. Fenton and Dr. Drinkell, who were involved in the interrogation of prisoners, it mainly concerns his recollections of the words and deeds of the Yorkshire Jacobite, Dr. John Burton. Bracken had met him in the street during the arrival of the rebels and Burton had told him that he had been captured by the Scots at Hornby and that he simply wanted to assess the numbers of the rebels so that he could inform General Wade. Bracken had then lost sight of him as he was summoned to the Cross, as a member of the Common Council, to proclaim the Pretender. At the next assizes, Bracken was released on a recognizance of £500.[14]

Whatever the truth of his activities, Bracken was rapidly rehabilitated, perhaps as a result of sympathy for his sufferings. As the second capital burgess, after Dr. Drinkell, he was elected to fill a vacancy among the aldermen in February 1746. Having taken the oath of abjuration, against James Edward Stuart, he was elected Mayor for the first time in October 1747 and for the second time in October 1757.[15] Bracken seems to have become more law-abiding, whether because his temper was less violent or his enemies were less active, and he hardly appears in court records hereafter. He was summoned in 1748 to answer charges that he had forcibly entered and seized a messuage in Lancaster in the possession of one Robert Deardon, to which he pleaded guilty and was fined. This was probably a dispute about the Mayor's rights of access as the Lancaster Corporation voted to allow £5 "out of the Corporation Stock to Doctor Bracken for the Charges of a Bill of Indictment lately preferred agt him for a Forcible Entry into the Town's Dog Kennell."[16]

Despite its relative calm, this period saw Bracken's most public dispute concerned purely with medical practice. It dealt with problems of skill and status and Bracken's comments epitomize not only his own previously expressed attitudes but also those of many other practical medical men when confronted with youngsters armed with the latest theories. Unfortunately for the historian, the scruples of Bracken's biographer prevented any discussion of "the paper-war that was carried on between him and Dr. Christopherson in the year 1747/8" ([6] p. 102). Since the pamphlet by Bracken's opponent seems to be lost, the dispute is only known from Bracken's side.

Bracken's printed letter to Christopherson, dated 10 Feb. 1747, is known from a nineteenth-century transcript which appears virtually complete (for a reprint of this, see [12] pp. 161–164). Christopherson had taken exception to a paragraph published in the *St. James Evening Post* of Dublin about one of his cases and had circulated an outraged pamphlet, accusing Bracken of slandering him. Since he was a Whig common councillor, whereas Bracken was the Tory Mayor-elect, he was probably preaching to the converted. In his reply, Bracken is incensed since for a recently graduated Bachelor of Medicine, "such a junior, to pretend to call me to account, is very odd, for I am ready to shew that, although you may be a Fellow of Pembroke Hall, yet you are but a silly Fellow, by reason that your letter is full of little else besides vain boasting and self-conceited ignorance."

Bracken discusses four cases which he argues Christopherson blatantly misdiagnosed. The first involves a man killed as a result of a fall from his horse, whom, Bracken contends might have been saved by trepanning. Christopherson and Dr. George Carlisle of Kendal, who attended the man, however, knew nothing of surgery and so the man lost his life. The second case involved the apothecary-merchant Thomas Butterfield, a Whig capital burgess, "a person for whom I had a great esteem, notwithstanding Party divisions, and disputes about Elections, had prevented that intimacy that otherwise might have subsisted between us." When Bracken saw him, he "was in the last stage of a dropsy, and by your directions had drunk such large quantities of cold water, along with a course of soapy medicines, that it was not possible to set him to rights." In this case, a suspicion of conspiracy enters the scene. Apparently, four physicians were in attendance and Christopherson accused Bracken of slighting the labours of his betters. It may be that Christopherson was acting as the cat's-paw of a more senior man, such as Dr. Drinkell, who did not care to enter the fray.

The other two cases involved dropsical women, one of whom had been seen by Christopherson for months without having her ailment identified. Bracken advised her to consult Thomas White in Manchester, since she was going that way. The second case involved Mrs. Haresnape in some way, and Bracken suggested that, although no midwife could give an adequate clinical description, she knew the patient's case rather better than Christopherson did.

Finally, Bracken responds to his adversary's boasts of the quality of his education by describing in detail his own, in Wigan and London, in Paris and Leyden, laying emphasis not only on the superior advantages of his experiences over any to be had in Cambridge but also on the very *practical nature of the skills that he acquired.* He is thus able to turn back the suggestion of being a mere tradesman onto Christopherson: "you make the noble Art of Healing more a Trade than a Science, and consequently you prescribe more for your own and the Apothecary's gain than the Patient's health, otherwise you have no occasion to dangle so much after people in imaginary illnesses." This is the nub of Bracken's medical ethics, based on the example of his mentors, rather than any religious or philosophical creed.

Although Bracken was a lifelong Church Tory, there is no trace of religious piety in any of his writings or during the greater part of his life. The most his biographer can say is that "he did not affect to talk much on religious subjects," but he appears to have become markedly pious during his prolonged final illness. We are told that "during the long confinement that preceded his death, every day, nay almost every hour of every day, was marked with some serious and pious act. Indeed the vanities of this life were then totally swept from his mind, and the inquisitive spirit that he still possessed was bent only on contemplations that had a reference to futurity" ([6] pp. 102–103). Religious devotion appears to have replaced his enthusiasm for horses.

Towards the end of his life, bodily frailty seems to have put almost a complete end to his beloved riding. In 1760, he had rented a piece of land in the Castle Ditch from the Corporation.[17] According to a codicil to his will, witnessed in December 1763 by William Moss, he had built a coach house on the land. In the preamble to his will, which he wrote himself a year earlier, Bracken testifies to his own weakness: "In the name of God Amen. I Henry Bracken of Lancaster in the County of Lancaster doctor in phisic being much afflicted with pain and Weak in body and fully convinced from a long experience in my profession that the time of our Mortal Life is exceedingly precarious and uncertain therefore I make and ordain this as my last Will and Testament."

His body was to be buried near those of his children, without much expense, "as I have always disrelish'd Pomp and Pageantry on these Occasions." As for his soul, "I surrender it into the protection of that great and omnipotent God who first animated my human Frame and preserves the Vital Spark throughout this Life." He confidently looks forward to being raised immortal, through the merits and passion of Christ, and to joining the heavenly choir. His cows, horses, and husbandry gear were to be sold to pay debts and his assets were to be turned into an annuity for his wife, "as I have never had reason to charge her with being a bad Oeconomist."[18] He had been under some anxiety about provision for his wife and apparently managed to raise just enough before he died for an adequate annuity to be purchased. As William Moss platitudinously commented, "The drama of his life was certainly closed most creditably for himself, as well as comfortably and respectably for his relict" ([6] pp. 102–103).

Having seen off our physician to his deserved rest, it only remains to assess what we know of his behavior, as it relates to medical ethics. To what extent did his quarrelsomeness affect his medical practice? Clearly, he did not aspire to emulating the tradition of "sad, grave physicians" advocated by Galenists in the previous century and to some extent continued by nonconformists in his own era. Nor did he model himself on the fashionable London practitioners, those bland men of fashion with their science and their antiquities. If anything it was the Tory squire who was Bracken's model, despite his failure to ever purchase an estate to which he could retire. All his pastimes were those of the rustic squire, from hunting and fishing to spending an evening in the tavern. Since it was among just such people that he sought his clientele, the scandals that Whigs attempted to fix on him are unlikely to have had any serious effect on his regular patients. The scandals did, however, prevent him from moving to London where rumor would soon have damned him and perhaps prevented him from establishing a really prosperous non-partisan practice in Lancaster and its hinterland.

Did Bracken deliberately deceive his readers into thinking him a graduate? Doubt has been thrown on the belief of earlier writers that Bracken possessed a medical degree, since he never claimed to have one, despite his references to Boerhaave, which might be calculated to mislead the unwary. It is frequently asserted by historians that physicians in the eighteenth century were always possessed of a medical degree, however nominal, although the example of David Hartley's belated degree makes it obvious that such a rule was far from universal. General practitioners in the eighteenth century were usually surgeon-apothecaries, but men like Bracken and Thomas White of Manchester exemplify an earlier type, the surgeon-physician. This partly explains the confusion about his status, as well as his own anxieties about young graduates. In most documents, Bracken was described as "gentleman" or "surgeon." The visitation of the Archdeaconry of Richmond in 1728, for example, listed him as "chirurgus" and that of 1733 called him "chirurgeon."[19] Bracken described himself in various ways but his publishers always describe him as "M.D." and he himself occasionally uses some equivalent notation. There was perhaps a slight element of defiance but it is likely that Bracken regarded himself as a "doctor of physic," since he practiced that part of medicine among others, and used the title fairly innocently since it would not deceive anyone who knew him. Later writers have been misled, however, into misunderstanding his disputes which arise in part from a typical eighteenth-century ethical issue, the role of qualifications and relative status [12].

Did Bracken exploit his apprentices? In view of Elletson's allegations, it would be interesting to know how Bracken treated his other apprentices. Apprenticeship was the usual method of training for general practitioners in the eighteenth century and even such luminaries as Fothergill started their careers as apprentices ([16]; [17] pp. 29–53). According to his biographer, "he generally had two or three at a time, each for about three years." He is said to have followed Worthington's example and instructed them diligently. "On this account the young men he sent into the world were soon found to possess superior qualifications; which, with the sanction of their master's celebrity, was a sufficient recommendation to them wherever they chose to offer their services to the public" ([6] p. 103).

Unfortunately, it is impossible to know even how many apprentices

Bracken taught as he only paid tax on four apprenticeship fees in 1747, two in 1754 and 1757. Since Bracken is admitted by his biographer to have been engaged in smuggling liquor from the Isle of Man, he may not have been an enthusiastic taxpayer ([6] p. 102). The fees paid for apprentices on these four occasions ranged from the £60 paid for James Geldart to the £32 paid for training Francis Atkinson for two years.[20] Two testimonials for earlier apprentices survive in the ecclesiastical archives at Chester. One is for the Cockermouth surgeon, Wilfrid Clementson, who studied at St.Thomas's after leaving Lancaster. It was signed by Bracken, described as "Chirurgeon," and Marmaduke Drinkell, MD, in 1731. The other was for Ralph Holt of Liverpool in 1755. Bracken stated that he had instructed Holt for five years "in the buisnesses I do profess (that is to say) as a Surgeon, Physician, and manMidwife."[21]

Apart from Holt and Clementson, who had reasonably successful practices, and the celebrated William Moss, it is difficult to know how well educated Bracken's apprentices were and what became of them. Against the success stories, one should perhaps counterpose a cautionary tale. James Geldart became apprenticed to Bracken in 1752, for four years. The tax was paid two years later.[22] As with the rest of Bracken's apprentices, his indentures were not registered in the borough records but, in any case, four years would not be long enough to secure the freedom of the borough. Bracken was careful not to encourage competition and did not take local apprentices who could obtain their freedom by patrimony. His main surgical rival was James Dickson, who had trained in the London hospitals after serving an apprenticeship to the Lancaster apothecary, Thomas Butterfield ([18] pp. 235–236).

When Geldart finished his time, Bracken was one of the assessors of the fine for his admission as a freeman. He paid five guineas to become the only one of Bracken's apprentices entitled to practice in Lancaster.[23] He appears to have been unable to compete with his former master and plied his trade among the rural poor. According to an indictment laid before the March, 1760 assizes, Geldart administered a pill the previous Christmas to a woman at Hornby in an attempt to induce an abortion. When this failed, Geldart and a slater assaulted her "and did then and there beat wound and ill treat the said Elizabeth Rigg so that her Life was greatly dispaired of with an Intent that the said Elizabeth being then and there pregnant with a Child or Fetus might bring forth the same dead...." Procuring an abortion was, at this date, only a misdemeanour, so Geldart was imprisoned for a year and bound over to be of good behavior for five years in the sum of £500.[24] Bracken had insufficiently instructed Geldart in both skills and ethics.

How did Bracken treat his patients? He certainly believed that most doctors were too considerate but little is known of his own relations with patients, apart from what he himself tells his readers. He repeatedly stressed surgical effectiveness rather than the blandishments of fashionable physicians. Against his glowing self-portrait, it should be noted that early in his career he appears to have pursued his fees more vigorously than was generally considered appropriate or decorous, and even imprisoned patients for debt. In 1725, Thomas Mashiter of Overton petitioned the Lancashire Quarter Sessions for relief. The father of four children, he had been sick and maintained by the township of Overton but he had been imprisoned in Lancaster Castle "at the suite of Henry Bracken Dr of Physick for a Debt unknown & not contracted by your Petitioner." He was supported in his petition by the curate, overseer, churchwarden and four inhabitants of Overton.[25] Even his biographer does not accuse Bracken of excessive charity, except to his friends.

He does, however, charge Bracken with indulging his vanity at the expense of patients: "In the time of the Doctor, and in country places in particular, it was too much the custom of the Faculty, when a patient's case was critical, or become hopeless, to foretell, out of the family, how he would go on, or how and when he would die, &c. To the display of this vain, and often cruel, kind of prescience he was greatly inclined, and indeed was perhaps, seldom excelled in the accuracy of such predictions." He attributes the decline of such prognostication since Bracken's day to "our progress in feeling and refinement." He also condemns Bracken for indulging people who came to him with a urinal for diagnosis. Bracken apparently believed that "Ignorant people should be dealt with a good deal in their own way." Although the biographer acknowledges the psychological aspect of this approach, he condemns it as "quackish." William Moss adds a footnote admitting that the doctor preferred to prescribe on the basis of urine "rather than from the blundering accounts of messengers" but he always ridiculed empirics who cast urine ([6] p. 102). These two aspects of Bracken's practice are open to various interpretations but he was clearly impressed with his own superiority even if he did not set out to preach enlightenment.

In what then did Bracken's medical ethics consist? For him, it was not genteel manners or an academic qualification that made a good doctor, but experience and care for the patient's welfare, even over-riding the patient's wishes for a painless therapy. He saw much of the medicine around him as fraudulent, aimed more at lining the doctor's pocket than at securing the patient's health. Bracken was as ready to criticize regularly trained practitioners, guilty of incompetence, as to attack boastful quacks and unskilled midwives. Above all, Bracken's career indicates the highly *political* nature of medical ethics, then as now.

REFERENCES

1. Bossy, J. (ed.): 1983, *Disputes and Settlements: Law and Human Relations in the West*, Cambridge University Press, Cambridge.
2. Bracken, H.: 1737, *Farriery Improved: or a Compleat Treatise upon the Art of Farriery*, London.
3. Bracken, H.: 1733, *The Gentleman's Pocket-Farrier*, Dublin.
4. Bracken, H.: 1739, *Lithiasis Anglicana: or, a Philosophical Enquiry into the Nature and Origin of the Stone...*, London.
5. Bracken, H.: 1737, *The Midwife's Companion*, London.
6. C[lementson], W.: 1804, "Some Account of Henry Bracken, M.D. late of Lancaster", *European Magazine* 45, 26–30, 100–4, 176–81.
7. Davies, W., and Fouracre, P. (eds.), 1986, *The Settlement of Disputes in Early Medieval Europe*, Cambridge University Press, Cambridge.
8. *Dictionary of National Biography*, entry for Henry Bracken.
9. Fildes, V.: 1986, *Breasts, Bottles and Babies: a History of Infant Feeding*, Edinburgh University Press, Edinburgh.
10. Fildes, V.: 1988, *Wet Nursing: A History from Antiquity to the Present*, Blackwell, Oxford.
11. Forbes, T.: 1971, "The Regulation of English Midwives in the Eighteenth and Nineteenth Centuries", *Medical History* 15, 352–62.
12. Harley, D.: 1990, "Honour and Property: the Structure of Professional Disputes in Eighteenth-Century English Medicine", in French, R., Cunningham, A. (eds.), *The Medical Enlightenment of the Eighteenth Century*, Cambridge University Press, Cambridge, pp. 138–164.
13. Harley, D.: forthcoming, "The Scope of Legal Medicine in Lancashire and Cheshire, 1660–1760".
14. Hartley, D.: 1739, *A View of the Present Evidence for and against Mrs. Stephens's Medicines*, London.
15. [Kennedy, P.]: 1739, *A Supplement to Kennedy's Opthalmographia; or, Treatise of the Eye; In which is observ'd The Plagiarism (from that Treatise) contain'd in Dr. Bracken's Farriery*, London.
16. Lane, J.: 1985, "The Role of Apprenticeship in Eighteenth-Century Medical Education in England", in Bynum, W. F., Porter, R. (eds.), *William Hunter and the Eighteenth-Century Medical World*, Cambridge University Press, Cambridge, pp. 57–103.

17. Loudon, I.: 1986, *Medical Care and the General Practitioner, 1750–1850,* Oxford University Press, Oxford.
18. Marshall, J. D. (ed.): 1967, *The Autobiography of William Stout of Lancaster, 1665–1752,* Chetham Society, Manchester.
19. McLaren, A.: 1984, *Reproductive Rituals,* Methuen, London.
20. Moss, W.: 1784, *A Familiar Medical Survey of Liverpool,* Liverpool.
21. Pitcarne, O.: 1739, *The Truth unvail'd for the Publick Good, or a Treatise on the Stone,* London.
22. Porter, R.: 1987, "A Touch of Danger: the Man-Midwife as Sexual Predator", in Rousseau, G. S., Porter, R. (eds.), *Sexual Underworlds of the Enlightenment,* Manchester University Press, Manchester, pp. 206–32.
23. Viseltear, A. J.: 1968, "Joanna Stephens and the Eighteenth-Century Lithontriptics", *Bulletin of the History of Medicine* 42, 199–220.
24. Wilson, A.: 1985, "William Hunter and the Varieties of Man-Midwifery", in Bynum, W. F., Porter, R. (eds.), *William Hunter and the Eighteenth-Century Medical World,* Cambridge University Press, Cambridge, pp. 343–69.

PUBLIC DOCUMENTS

[1] Lancaster Borough Minutes and Accounts (LBM) 1661–1736, p. 334; Public Record Office (PRO): PL 28/1, f.154r; PL 26/290, inquest on the male bastard of Mary Parker, 22 Aug. 1748; PL 28/2, p. 73.

[2] Annals of the Royal College of Physicians, 3 June 1737 & 5 August 1737.

[3] LBM 1661–1736, pp. 305–34; PRO: PL 11/24, pp. 305, 309, 319, 325, 330.

[4] Lancaster City Library (LCL): Plaint Books: MS 107 (7 June 1734, 19 Sept.1734, 1 Sept.1737); MS 103 (22 Oct. 1736, 28 July 1737, 16 Feb. 1737/8, 4 May 1738, 18 May 1738, 31 Jan. 1739/40, 8 Aug.1745, 3 Sept. 1747); MS 222 (8 Aug. 1745, 4 June 1747).

[5] PRO: PL 28/1, f.185r.

[6] LBM 1736–56, p. 12.

[7] PRO: PL 27/2: depositions of 24 Dec. 1736 & 3 Jan. 1736/7; PL 25/99; PL 28/1, ff.177–8.

[8] PRO: PL 25/102; PL 28/1, ff.180r, 183v, 184v, 185r, 190v, 191v.

[9] PRO: PL 28/13, pp. 10, 17, 29, 31, 36; PL 25/99; PL 28/1, f.178r.

[10] PRO: PL 11/24, pp. 425, 427, 448, 452, 459, 462; PL 11/18, 5 Aug. 1737, 23 April 1739, 26 Aug. 1740.

[11] PRO: KB 1/6, pt.1, Easter 13 Geo.2; KB 1/6, pt.2, Easter 13 Geo.2.

[12] PRO: KB 1/7, Trinity 16 Geo.2.

[13] PRO: C 33/397, p. 68; C/11/2529/26; PL 11/18, 2 July 1760 & 26 Sept 1760.

[14] PRO: KB 33/4/1; PL 28/2, p. 38.

[15] LBM 1736–56, pp. 44, 53, 55; LBM 1756–94, p. 10; LCL: MS 221 (7 May 1747).

[16] PRO: PL 28/13, p. 133; PL 28/2, pp. 60, 65; LBM 1736–56, p. 67.

[17] LCL: MS 160, p. 87.

[18] PRO: PROB 10/2401 (March 1765); PROB 11/906, f.303r–304r.

[19] Lancashire Record Office (LRO): DRCh 11, f.3v; DRCh 12, f.4v.

[20] PRO: IR1/18/52; IR1/52/25 & 30; IR1/21/103.

[21] Cheshire Record Office: Dioc. Misc. bundle 2, item 6; bundle 3, item 118.
[22] PRO: IR1/52/30.
[23] LCL: MS 160, p. 44.
[24] PRO: PL 28/10, ff.10v, 11r, 15, 36v; PL 28/13, p. 280; PL 25/141; PL 28/2, pp. 248, 250, 255.
[25] LRO: QSP 1235/2

CHAPTER 3

ROY PORTER

PLUTUS OR HYGEIA? THOMAS BEDDOES AND THE CRISIS OF MEDICAL ETHICS IN BRITAIN AT THE TURN OF THE NINETEENTH CENTURY

There is a history of medical ethics which rightly should be the history of theories, codes, and formal treatises. There is also a history to be written of medical cases – legal and administrative – which raise ethical issues. But there is a third kind of subject matter which forms part of the history of medical ethics: the analysis of wider reflections upon the nature and standing of the medical profession within society. Thomas Beddoes (1760–1808), the subject of the following essay, was a practical physician who wrote widely about politics, including the politics of the medical profession. He nowhere tried to codify medical ethics, nor did he reflect upon the codes of medical ethics which were beginning to circulate in his day. One may guess that he would have found them so much hot air, mystifying ideology irrelevant to the real dilemmas of the conduct of medicine in society. But he has a place in the history of medical ethics, since he reflected intensely and extensively upon the ethical basis – or its lack! – for the practice of medicine at the end of the old regime.

Reform was on the agenda for many British medical men around the turn of the nineteenth century, and the rhetoric they used to promote the cause was that of morality. The body medical had grown corrupt. There was a surplus, or, perhaps, a shortage, of practitioners; too many doctors were too old, or, maybe, too young, or undertrained, or over-qualified; some were too opulent, others too poor; and anyway practice was infested with a 'vile race of quacks', impostors, 'race of quacks', impostors, empirics and charlatans, mountebanks and itinerants. Much of the blame for this was to be laid squarely upon the shoulders of the sick – for playing fast and loose with poor, honest, long-suffering practitioners – but the responsibility for remedying this parlous situation had to rest with the profession itself, which needed to put its own house into order, probably with the aid of Parliament [56]; [41].

This moral rhetoric of reform, couched in the vocabulary of civic humanist virtue, is today familiar thanks to the researches of Irvine Loudon, Ivan Waddington, and others [38]; [39]; [40]; [76]; [77]; [29];

R. Baker, Dorothy Porter and Roy Porter (eds), The Codification of Medical Morality, 73–91.

[30]; [49]; [51. It was a summons to action which flowed easily and often from the pen of Thomas Beddoes, the turn-of-the-century Bristol practitioner who had quit teaching chemistry at Oxford University in 1793, partly because of antipathy to his radical politics, moving to found a Pneumatic Institution, finally established in 1799, at Clifton, just outside Bristol, where he passed the rest of his career in private practice, promoting the cause of health and threshing around in medical politics [73]; [71]; [10]; [17]; [20]; [33]; [35]; [36]; [37]. Often prolix and hectoring, sometimes needle-sharp, ever the earnest moralist, though capable of flashes of caustic and even surreal wit, Beddoes exposed the pathology, indeed the psychopathology, of the body medical in a torrent of publications from the early 1790s untill his death in 1808 [59].

What makes his reflections upon medical ethics – past, present and future, descriptive and prescriptive – so memorable is that he sidestepped the "dearly beloved" pieties employed by other such adepts of the genre as Gregory and Percival [12]; [50]; [75], and transcended their benevolist vision of "the achievement by physic of a more dignified *esprit de corps* while, at the same time, better serving the public." Beddoes thought the ills of medicine systemic, constitutional, terminal even. Many reformers of the 1800s looked to change through tinkering; Beddoes espoused the radical philosophy of root-and-branch [74]; [19]; [13].

For the malaise, in Beddoes' view, was not just a question of collegial corruption, oligarchic blight, and charlatan voracity; rather, it was structural. The fundamental problem was quackery. Like his peers, Beddoes hated quackery with all his heart, deploring the decimation of poor people by ignorant-hucksters, profiteering hand-over-fist out of the ever-gullible-public, and wrecking the nation's health in the bargain [41]; [55]; [11]. Time and again he warned his lay readers against saturation – advertised nostrums, such as mercurial vermifuges for infants, which he regarded as little short of rank poison. "In consequence of money expended on their purchase," he fumed, "families above the poorest class are frequently deprived of the necessaries of life .... There are instances where people have sold the bed from under them (the rage for quack medicines, in some familiar instances, being just the habit of dram-drinking)" ([5], p. 98).

More incensed, more optimistic, maybe more dictatorial than his colleagues, Beddoes was for public action against such quacks. Bodies should be established – presumably like the Société Royale de Médecine – to assay proprietary preparations; and, not least, bare-faced and impenitent quacks should be outlawed by legislation: "A scheme for the reform of medicine without the abolition of quack medicines, is about as hopeful as one for making the rattle-snake harmless" ([5], p. 98; [61]; [62]; [63]). It was no time for compromise. "The advertising and sale of secret medicines ought to be entirely suppressed" ([5], p. 99).

Fulminations against quack doctors were ten-a-penny in the medical press of the day; they need no elaboration here. The true radicalism of Beddoes's attack lies in his indictment against the profession as a whole, that it was itself practicing not medicine but quackery – a palpable hit scored already, of course, by who knows how many Grub Street satirists [18], but rare enough as a self-confession from the pen of an august Oxford M.D. Time and again, in his syllabus of medical errors, Beddoes pinpoints the parallels between quacks and regulars – or "medical Jesuits" and "medical Jansenists," as this ferocious foe of priestcraft was pleased to dub them ([5], p. 105).

Regulars vilified empirics, yet was this not a barefaced case of the pot calling the kettle black? Quacks were lambasted for advertising and self-publicizing. But, begged Beddoes, did not they, in so doing, merely "manifest the spirit of the trade" ([5], p. 108) – for surely regulars were no laggards in touting for clients, albeit behind a veneer of gentility: "Indeed to coquet for custom in an equipage beyond his circumstances is considered as altogether essential to his success" ([5], p. 106). In the same way, the novice practitioner was always "advised to give so many *dinners of business* per annum," a tactic whereby he gets his wife – his "serjeant Kite in petticoats" – to "recruit for patients." "The good lady crams them with her dainties, in sure and certain hopes, that the turn of her doctor will come to cram them equally with his drugs." So where lay the essential difference between regulars and their "bastard brethren?" ([5], p. 107) Therapeutically and pharmaceutically they were like as two peas in a pod. "Of quack compositions we regulars cannot in honesty but confess that they are excellent, being, in fact, the very same which we use ourselves" ([5], p. 130; [53]).

This device of superimposing regulars and quacks as doubles was Beddoes's graphic demonstration that practitioners had descended into practicing healing as a branch of commerce, rather than as an exalted ethical calling. Disregarding lofty ideals, modern physicians, unlike, perhaps those of the Humanist era, had become hagglers in the market place, their sights trained not upon health but wealth [78].

Modern medical mercenariness hardly bothered to wear a fig-leaf.

"Physicians are often as needlessly prodigal of their visits," Beddoes accused, "as apothecaries of their potions" – a damning comparison indeed ([5], p. 110). A *bon mot* of the salon: "There is no getting doctors and workmen out of a house, once they get in" ([5], p. 110]). Once upon a time, physicians had cultivated the arduous art of regimen. Gone were the days. Now they treated only by drugs, reducing healing to a commerce based on commodities. "The public has learned to expect scarce any thing beyond particular prescriptions; and to minds little inured to reasoning, scarce any thing else is acceptable" ([3], vol. 1, essay 2, p. 48). And was not this profitable simplification of the *ars medendi*, the product of a notorious unholy alliance "between doctor and apothecary," or rather "drench dealers" and vendors of "medicated wash?" ([5], p. 108]) "The desire of the apothecary to swell his bill, and the complaisance of the physician towards the apothecary," Beddoes regretted, "are justly believed to create infinite abuse in our profession" ([3], vol. 2, essay viii, p. 75).

Survey the facts of medical practice. Did one find that orderly tripartite, incorporated hierarchy, College of Physicians at the apex, beloved of traditional medical historians? Not a bit of it. There was a sething mass of heteroclite healing activity ([5], p. 96]), in which it was quite impossible to draw the dividing lines, separating the sheep from the goats, men of integrity from quacks. Certainly, popularity and reputation would never serve as a litmus test:

> Without going a hundred miles from Clifton, Bristol and Bath, you may meet with practitioners, whose genius has transported them at a single bound from the side of the mortar to the bedside; and who go about distributing their *poudres de succession* through town and country, with as much professional gravity, as if they had gone through the longest course of study, and stood the severest trials of skill. As to consumption-doctresses, cancer-curers, mechanics professing to treat divers disorders, and particularly those of the female sex, there have arisen within my short memory, several, in whose behalf to speak with the cricket players, one might safely challenge all England; nay, in this favoured district, do we not behold the splendid seats of solemn, regular, respected quackery, methodized, as you shall hear, if you do not know it, according to the most approved forms of a foreign merchant's counting house ([5], p. 9).

Regulars too owed their own rise in clientele and credit to the world's applause, which recognized fortune as the sole yardstick of worth. "Our dignity," Beddoes bewailed, "is unfortunately placed in the quantity of our gains, not of the good we do" ([5], p. 111; [46]; [1]; [15]).

Gold had become the *primum mobile*, the *summum bonum*, the very life-force of medicine. "Money, I perceive, can put all the members of the faculty and all its appendages into busy motion" ([3], vol. 1, essay i, p. 72). Hence medics had oriented their performance to the demands of gain. "But to what purpose? Very frequently, not to the purpose of deferring, one hour, the fatal crisis. If indeed, all this bustle console the sick, or the survivors, then must it be considered as well purchased, at whatever price." But in truth, "the doctor's pomp and apothecary's mysterious hurry" actually achieved nothing of therapeutic value, "so impracticable has it been found to reduce health to the state of a mere article of commerce, and so entirely does HYGEIA disdain to become the slave of PLUTUS" ([3], vol. 1, essay i, p. 73).

Beddoes was not above hurling an occasional *"j'accuse"* at specific practitioners whose love of lucre was grotesque. "I have been assured, by a curious observer," he informed Sir Joseph Banks, "that the late Dr. Warren often bestowed but three minutes on a case. I understand that a fashionable physician in town is not to bestow above ten minutes under the peril of being deemed not sufficiently hurried" – Richard Warren was so obsessed with lucre, it was said, that as he inspected his tongue in the mirror in the morning, he automatically transferred a guinea from one pocket to another ([5], p. 128). But Beddoes's prime aim was to lay bare the structural features promoting and perpetuating this perversion of medicine into what he dubbed 'the sick trade' ([5], p. 100).

Beddoes was a trenchant critic of the ascendancy of commercial capitalism:

> In the social arrangements which have gradually formed themselves in Europe, WEALTH, the most general object of power, becomes the most general object of desire .... The multiplication of the roads to wealth unavoidably keeps pace with the multiplication of gratifications, and of those contrivances for displaying accomplishments, which wealth can command; for it is by profiting from the sale of these gratifications and these contrivances (that is, of the various luxuries of the table, of the toilet, of furniture, of equipage, of the fine arts) that more and more members of the society grow rich ([3], vol. 1, essay ii, p. 54).

Were the dynamics of the birth of the consumer society ever more lucidly, more succinctly, stated? A "chain of destructive vanity" ([3], vol. 1, essay ii, p. 62) binds class and class into this cash nexus mechanism, whose wheels were driven by that "fawning, treacherous divinity," fashion ([3], vol. 1, essay ii, p. 62). Fashion in turn, of course, made work for the medics, for the wealth of nations inevitably sapped the health of nations – "does not gold bring with it its plagues?" ([3], vol. 1, essay ii,

p. 57)

Surplus, disposable income had mushroomed; there was an intensification of getting and spending, the buying and selling of ever-multiplying consumer items. Not least, goods and services were undergoing a process of industrial commodification, being turned into mass-produced, standardized products each with its price in the market-place [43]; [44]; [66]; [14]; [28]; [42]. Beddoes contended that medical practitioners formed an integral part of this capitalist system, indeed, were cashing in on it. They had eagerly accommodated themselves to the commercial ethos governing a free exchange, laissez-faire market society. Since doctors traded like shopkeepers, might it be no bad idea, pondered Beddoes, with 'caveat emptor' in mind, if "medical certificates should have written on them CAVEAT AEGROTI?" ([5], p. 71)

The infection of doctors by the morals of the marketplace was bad enough. But there was a further facet to the problem, reflecting another feature of manufacturing society. The law of supply and demand was sovereign. Doctors had set themselves up as medical suppliers, traffickers in diagnoses and drugs. The demand-side of this equation was constituted by the sick – or frequently the pseudo-sick. Today's economic historians are acknowledging that industrialization was at least as much demand as supply-led [9]; [6]. I believe that such an interpretation would hold good also as a way of accounting for the dynamics of Georgian medical expansion. Beddoes, I suspect, would have agreed, for he routinely portrayed doctors as suppliers operating in a state of sycophantic subservience to their 'sick trade' customers and all their fads and footling foibles.

The sick person, thus runs Beddoes's damnation of demand-led bedside medicine, always knew best – and, as the paying piper who called the tune, exercised the power of the purse. The Quality plumed themselves upon their expertise in matters of medicine – they even cultivated the affectation of referring to themselves as "private practitioners," snarled Beddoes – "what a winning thing is a genteel name," being comparable to dubbing an assasin a "private gentleman soldier" ([3], p. vol. 1, essay ii, p. 20). Such "private practitioners," the worst offenders amongst whom were women, "busy-bodies in petticoats," presumed they could diagnose, prescribe, and generally consult with physicians as equals if not superiors ([5], p. 113).

Beddoes appreciated that the power-base for this patient control in medicine stemmed from the perdurable strength of personal patronage in the unreformed socio-political order – he would surely have nodded agreement with Nicholas Jewson's recent Namierite account of the cognitive politics of the pre-reform medical profession [26]; [27]. Grandees expected deference from their physicians much as they did from all underlings, from their tenantry and political agents down to butchers, bakers and candlestickmakers. Thanks to aristocratic hegemony, lesser gentlefolks were able to emulate their manners. At bottom, therefore, doctors operated as no better than high-grade flunkeys in an grandee service economy – one in which the paymasters had not the slightest reason for treating regulars preferentially to quacks (where, once more, lay the difference?). The names of peers and parsons, Beddoes observed, were always to be found "dangling by dozens, to the tail of frauds at half a guinea the bottle," in newspaper quack bills ([3], vol. 1, essay ii, p. 16).

Practitioners were obviously as impotent to abolish the patronage system at large, as they were to rescind the action-at-a-distance exerted by cash. They had, however, Beddoes regretted, needlessly made a bad situation worse. In two ways in particular.

First, the profession was to blame for positively encouraging polite society to entertain grossly exaggerated – and utterly destructive – conceptions of lay medical capabilities. One increasing source of this evil lay in the torrent of "Every Man His Own Physician" tomes flooding off the presses, all too many of which were the handiwork of regulars, thereby providing yet a further instance – this time à la Grub Street – of the degradation of medicine into an article of commerce ([69]; [70]; [32]; [67]; [57]. (In making such indictments, Beddoes was naturally inviting ripostes concerning the right to cast the first stone, since there was no more energetic penner of pamphlets than Thomas Beddoes; but he believed a radical distinction could, and should, be drawn between profaning medicine before the laity by pretending to expound the science of drugs and cures, and (on the other hand) his own high-minded mission of openly displaying the tablets of the laws of positive health and prevention [56].)

By thus encouraging the laity to think themselves medically capable, expert even, physicians had heaped coals upon their own head. For, through such books, "opinions cast off, like threadbare apparel, by the faculty, go each in their turn to the public." Popular treatises spread diseases ([3], vol. 1, essay i, p. 5). "*In the politest and best informed circles*, according to newspaper phraseology, no day probably passes without its victim to some antiquated hypothesis concerning the *purity of the blood, the acrimony of the juices, the transmigration of humours, the salubrity of the air of this or that spot*" ([3], vol. i, essay i, p. 5). Thanks to a rum sort of "Error's Progress," "many suppositions reign for a time in the schools, become exploded, and afterwards make their fortune in the world" – with devastating effect ([3], vol. 1, essay i, p. 52; [58]).

Some such bookish merchandize had become appallingly popular. "Certain adventurers," Beddoes names no names, but one surmises he may be referring to William Buchan, or even, amongst an earlier generation, to George Cheyne, "have obtained a degree of public confidence by their endeavours to popularize the practice of medicine" ([3], vol. 1, essay ii, p. 33; [60]). Yet all such books were dross, and must remain so – *subsequent* experience has not enabled them to advance a single step towards conquering the impossibilities inherent in the undertaking ([3], vol. 1, essay p. 33). This prostitution of medicine to market forces by "the projector of a new domestic medicine" was "a scheme perfectly in the spirit of our literary traffick." "To him, who has only in view the making of money, a grammar, a gazetteer, a medical compendium will appear identical, provided they prove equally saleable articles" ([3], vol. 1, essay ii, p. 35).

On the whole, Beddoes concluded, do-it-yourself medical texts were more lethal than highwaymen: gentlemen of the road at least left victims a choice between "your money or your life." With books of auto-medication, by contrast, the rule was your money *and* your life. "Quacking books are unquestionably the same evil wholesale which quack medicines are by retail" ([4], p. 24).

With such books in their hands, or on their shelves, the sick, above all, the new brood of hypochondriacs, would pride themselves upon their erudition, take "up the terms in which medical opinion has been delivered, and use them as battledores to strike nonsense backwards and forwards like a shuttlecork" ([4], p. 26). The somber truth was that the public was ignorant about authentic medical practice, and should stay that way. Of course, there was endless canting about the sovereign voice of public opinion, but what had that to do with medicine? "It is sometimes not impertinent to ask," Beddoes insisted, "if there be any sense in the *vox populi?*" ([4], p. 26)

Apologists might insinuate that free choice in the free market economy would result in the best doctors rising to the top by a kind of spontaneous public acclaim. But this was all baloney: informed choice was a myth – for who could deny that "a great part of this very public is incapable of distinguishing square from round, black from white, in the forms and colors of medical character" ([5], p. 26). By the criterion of public choice, "the usual signs of public confidence in medicine," the most illustrious physician in Britain would appear to be none other than the notorious Liverpool quack, Samuel Solomon, "a fellow who orders British gin from Bristol to Liverpool, colors and christens it balm ... and who enjoys as much of this confidence as almost all the fellows of the three royal colleges put together" ([5], p. 26; [55]). In short, demand from below, from a mis-educated public seeking bargains and diversions in the medical market-place, could do nothing but reduce medicine to whoredom and travesty.

Second, practitioners were further digging their own graves – or at least betraying their cloth – by their shameless acquiescence in abject toadyism, bowing and scraping to the jingle of guineas, for the ends of ingratiation and advancement. Beddoes was appalled at the obsequiousness of tuft-hunting courtier physicians such as Thomas Gisborne (who is not to be confused with the theologian and medical ethicist of the same name). Gisborne, the physician is described as a notorious sycophant "member of the *haute noblesse* of medicine." Beddoes relates the tale of "One of the Princesses being taken ill, and Dr. Gisborne in attendance":

> [H]er royal highness enquired of the doctor if she might not indulge in the use of a little ice cream, as she thought it would greatly refresh her. Dr. G. who never contradicted his royal patients, answered that he *'entirely agreed with her royal highness;'* and the ice was accordingly provided. His Majesty, visiting the chamber and observing the glass, with some of the ice still remaining in it, seemed alarmed, on the supposition that it might be improper; but her royal highness assured him that she had the doctor's permission for what she had done. His Majesty ordered the doctor into his presence, and observing to him that he had never heard of ice being recommended in such cases before, expressed his apprehension that it was on some new system. The doctor seemed at first a little confounded, but quickly recovering himself, replied, *'Oh no, please your Majesty, it may well be allowed provided it be taken warm'* – *'Oh well, well, doctor, very well, very warm ice, warm ice'* ([5], p. 115; [48].

"Thus," concluded Beddoes, "are the Pretensions of the vulgar and the stupid fostered by flatter" ([5], p. 116). And if the Court encouraged such servility, "the greatest proficients" in this line of business were the hireling physicians at spas and watering places, who never fell "short in the great talent of simpering and bowing," above all "our medical brethren from the other side of the Tweed," about whom more below ([4], p. 331).

Beddoes volunteered a recipe for the heapishness vital for a practitioner determined to get rich quick amongst the powerful. Serving such patients whose "minds are usually made quite up" as to diagnosis, prognosis, and treatment alike', the best thing which the doctor can do, Beddoes commented,

is to listen with a face of sanctified wonder, protest that he was just thinking of the same thing, put his goose-quill under the guidance of their inspirations, and content himself with translating them into certain magical abbreviations, as *pilul.- pulv.- haust.*, which are the main support of his professional dignity and that grand mystery of the art, into which these his female prompters have not yet penetrated ([5], p. 114).

Such complicity and rapacity amongst the "*petites maîtres* of physic that figure as favourites of the great" ([3], vol. 2, essay v, p. 66) must spell bad medicine, even if, Beddoes avows, disarmingly, "I would not be so rash as absolutely to affirm that complaisance destroys more lives than contagion" ([3], vol. 1, essay i, p. 71). This pantomime of deference had quite perverted physic, patient-led medical demand had got out of hand, and nowadays:

[T]he practitioner of physic is forced to prescribe for the alarms of patients, when there is no call upon him from their danger. To treat their false fears lightly would be attended with the certain effect of forfeiting their confidence. And then there would be a degree of danger lest they should find the terrible and sublime, which many a quack bill so happily blends with the soothing, more congenial to their feelings than unadorned truth ([3], vol. 1, essay 1, p. 60).

In short, patient patronage and purse power cast evil shadows over healing. Practitioners performed as tradesmen. Such, of course, presumably came naturally for hundreds, perhaps thousands, of journeyman surgeon-apothecaries out in the sticks, whose medical training through apprenticeship had never inclined them to think that they were following a calling superior to that of butcher or sow-gelder [30]; [8]. But it also applied higher up the tree, where courtier physicians gleefully acquiesced in aesculapian prostitution because they were doing very nicely out of it. Who could be surprised at the "lukewarmness towards improvement," Beddoes demanded, "which so commonly arises when the physician has got upon the full scent of profit, and still less of the apathy which overtakes him as soon as he 'has feathered his nest'?" ([3], vol. 1, essay i, p. 60)

Not that he felt entirely without qualms about the motives of some of the more vociferous promoters of reform. Their drive against quackery, and their insistence upon paper qualifications as the *sine qua non* of practice, seemed themselves to smack of a new spirit of monopoly. Might not reformist zeal be a rationalization of the "jealousy of intruders ([5], p. 9)" on the part of those "galled by their competition" – those seeking to appropriate exclusively to themselves the "privilege of lucrative homicide?" ([5], p. 97) So, if the watchword of the times was *"Physician, heal thyself,"* how then was this to be achieved? ([5], p. 11)

In Beddoes's view, medicine would become an ethical profession only by escaping the clutches, the seductions even, of trade. He entertained severe doubts as to whether some of the most touted reform proposals would actually accomplish this – or rather, by contrast, they would merely promote and perpetuate the commercial spirit perverting the profession. Take the question of medical education. One influential reform voice contended that all practitioners within a purified profession should have passed through at least a three-year medical training, rather on the model of that afforded by Edinburgh University [5]; [22].

Beddoes granted many virtues to the Edinburgh school [32]; [24]; [47]; [54]. For long, "our supine and almost criminal neglect of our own advantages suffered Edinburgh to continue almost the only place in the three kingdoms, where at once degrees were conferred and lectures systematically read" ([5], p. 55). Yet such were the "extravagant assumptions of Edinburgh" ([5], p. 56), broadcast by the university's own self-serving propaganda lobby, that it was easy to forget its crying evils, not least its nepotism, or what Beddoes called "the system of hereditary professorships" ([5], p. 79) – a system with "every reputed disadvantage of hereditary monarchy, and not one of its advantages" ([5], p. 78; [64]; [65]).

Edinburgh offered a poor model for medical education. For one thing, its three-year course was far too short. It was (and Beddoes here spoke from personal experience) all rush and cram, with hordes of students dashing from lecture to lecture – many of these superficial: "May Esculapius protect his votaries from ever again depending on *demonstrations* in the style of the elder Dr ***** [i.e., Monro]" [32]. Students had no time to read, but what matter? They could easily creep through the farcically stereotyped finals thanks to the ministrations of external coaches aptly called "grinders." All was thus mechanical; as a consequence, "do not blockheads enough pass through the Edinburgh sieve?" ([5], p. 38)

In short, in an industrial age, Edinburgh was the new model medical degree-factory *par excellence*, grinding the maximum number of students off the production line at minimal cost, anticipating the Bell and Lancaster system in elementary schools.

Because students went up to Edinburgh almost as schoolboys, and graduated while still striplings, the nation was being filled with "half-drilled medical recruits" – tyros too young, too inexperienced, to be fit to practice – though so what? – since it was easy enough for such a youth "to persuade some credulous knot of old ladies, that he come from Scotland full charged with healing virtue" ([5], p. 41). "I suspect," guessed Beddoes, "that a good judge of medical stock would find many physicians educated during the greatest splendour of the Edinburgh school ... among the sorriest sheep in the whole flock of Esculapius" ([5], p. 36). He was bound to get accused of illiberality, Beddoes confessed, yet he thought it no bad thing that the Royal College of Physicians cold-shouldered these medics on-the-move and on-the-make ([5], p. 39).

Medical education should not be organized on the factory system. More time was needed ([5], p. 59). Rather than the Caledonian "triennial manufactory," a full six years should be allotted for study ([5], p. 72). Seventy-two months would allow the student an evens chance to "digest from 500 to 800" volumes of medicine – such Beddoes judged the minimum fit to acquaint him with the art and science, to cultivate a well-stocked mind, and to permit a tempered, experienced clinical judgment to mature. Medical erudition was not to be despised. Who, after all, had been the greatest clinician of the previous generation? William Heberden, a soul 'singularly learned' ([5], p. 74; [23]). Heberden was, of course, a product of the Cambridge system. There was, assuredly, little enough to praise about the quality of contemporary Oxbridge medical teaching – did it even exist? – but Beddoes was an ardent enthusiast for the philosophy enshrined in the ancient universities, designed as it was to foster independence of judgment, a liberal spirit, and, finally, ensuring that the student ripened in years and character before embarking upon practice.

Not least – and one may be excused for thinking these remarks ring somewhat strangely, coming as they do from a sworn foe of place, privilege and snobbery – the promotion of a liberal, Oxbridge-style medical education would decisively attract into the profession men of rank, family and breeding, who would endow it with the backbone of authority. "Medicine," Beddoes deplored, "is the most servile of the professions, and still bears so strongly the marks of the ancient condition of its members. One has heard of fawning divines, fawning courtiers, and the like. But these spaniels in the human form have only to caper at the whistle of a single master; whereas the supple carcase of the physician must perform its cringes to a whole circle every day" ([5], p. 114; [79]; [16]). Why this degradation? It had not a little to do with the fact that medicine was entirely composed of "us plebeians" ([5], p. 50). If only more of the Quality were to enter medicine, as they perched on the higher branches of the Bar and the Church ([5], p. 49), and if only the Crown would further help by knighting or ennobling the top notch, the prestige of the profession would inevitably soar, and with that, its independence.

Thus Beddoes's alternative to Edinburgh's "too rapid manufacture of physicians" ([5], p. 52) was a revitalization of the Oxbridge educational philosophy, designed to confer upon medical graduates the intellectual stature accorded by a liberal education. He was confident that herein lay the remedy to commercialism, because, as an Oxford B.A. and M.D., he knew it from his own experience: "Having in early life been altogether a stranger to medicine as a trade," he confided, "I naturally acquired the habit of regarding it purely as a body of doctrine, productive of certain advantages to society, and as respectable or the contrary, in proportion to the amount of those advantages" ([3], vol. 1, essay i, p. 55).

Once medicine were peopled – or at least led and leavened – by practitioners of true intellectual stature, displaying the ethos of independent medico-scientific authority which accompanied it, proper independent clinical judgment and honest practice could reassert themselves. The cash nexus, which – polite pretences to the contrary – dominated medical practice, would wither away, to be replaced by medically appropriate and ethical doctor-patient relations. The grateful patient would defer to the physician's superior wisdom; the physician would address himself to medical needs, not to advancement. The danger was that medical reform would be sidetracked by endless memoranda about certificates, courses, examinations, degrees, portals of entry, and professional associations. All such matters were peripheral, in contrast to the need for medicine – if it were to be truly ethical – to stand outside and rise above the market imperatives of the commercial rat-race.

Medical ethics, in the normal, narrow sense of the code of conduct governing practitioners, mattered to Beddoes. It mattered because it was indispensable for something much more fundamental. For Beddoes believed that medicine itself – or at least the bio-medical understanding of life – was the cornerstone of ethics as a whole. Chemist, physiologist, and, perhaps, atheist, Beddoes was disposed, like his French contemporaries, the Idéologues, to take a materialist perspective upon the entire panorama of animated existence [72]; [45]. He ridiculed the attempts of Cartesian dualists to draw sharp divides between mind and brain, spirit and flesh, the material and immaterial – such sciolists overlooked "the inseparable connection between moral and medical topics" ([3], vol. 1, essay iii, p. 83). Through advancing numerous clinical instances of psychosomatic and somatopsychic complaints, Beddoes denied any rigid ontological differentiation between the physical and the moral. Hair-splitting metaphysicians he held in utter contempt. For all his admiration for *Sturm und Drang* culture, Beddoes rarely had a good word to say for the dualists.

No moral system – or, more broadly, no code of living, no value system, or even lifestyle – could hold any real value, could be *moral*, unless grounded upon, and conducive to, the permanent well-being of man as an organic being. Physiology was the basis, and test, of all sound morals, for "the science of human nature is altogether incapable of division into independent branches":

> Every code of morals must ground its precepts on a comprehensive view of the laws that regulate feeling, and deliver the conditions of an offensive and defensive league, having for its object the well-being of individuals. Without accurate ideas therefore of the causes that affect the personal condition of mankind, how is it possible to conceive any progress in genuine morality? And will not every addition to this branch of knowledge necessarily tend to purify morals – that is, to introduce into the social compact covenants more beneficial to the parties? Without reference to the body, it is equally impossible to unfold the nature of the mind. Physiology therefore – or, more strictly, *biology* – by which I mean the doctrine of the living system in all its states, appears to be the foundation of ethics and pneumatology ([2], p. 3).

Thus there was an ethics of medicine; but, more importantly still, medicine itself must be the yardstick for all possible ethics. The physician thereby became the legislator of the new moral world. The doctor's duty was to instruct the people in the lessons of living, because without understanding these, how could they possibly make the wider choices of life? It may be no accident that Beddoes was a close acquaintance of Samuel Taylor Coleridge [12]; [25]; [34]. Beddoes's vision of the doctor as intellectual guru parallels Coleridge's vision of the new intelligentsia, the clerisy, called to serve as moral leaders of the future. No servile tradesman, no sycophantic courtier, no technocratic fixer, the Beddoesian doctor would be the humanist physician – priest to humanity.

In Beddoes's lifetime, the great formulators of codes of medical ethics – Gregory, Percival – were concerned to find acceptable legitimations of the personal relations between patients and practitioners and, no less, between the various members of the medical profession itself. In his rather brusque way, Beddoes brought stage-center a more crucial issue. Medicine had to choose between the cash-nexus and the knowledge-nexus. Medicine could be moved by money, or it could be animated and organized by science, by the imperatives of the search for truth. However impractically, however unrealistically, Beddoes thereby articulated one of the most fundamental issues of medical ethics.

The Wellcome Institute for the History of Medicine
London

REFERENCES

1. Barry, J.: 1986, 'The Cultural Life of Bristol, 1640–1775', D. Phil. thesis, Oxford University.
2. Beddoes, T. (ed.): 1799, *Contributions to Physical and Medical Knowledge, Principally from the West of England*, T. N. Longman and O. Rees, London.
3. Beddoes, T.: 1802, 3 vols, *Hygeia: or Essays Moral and Medical, on the Causes Affecting the Personal State of our Middling and Affluent Classes*, J. Mills, Bristol.
4. Beddoes, T.: 1806, *Manual of Health: or, the Invalid Conducted Safely Throuqh the Seasons*, Johnson, London.
5. Beddoes, T.: 1808, *A Letter to the Right Honourable Sir Joseph Banks ... on the Causes and Removal of the Prevailing Discontents, Imperfections, and Abuses, in Medicine*, Richard Phillips, London.
6. Berg, M.: 1985, *The Age of Manufactures, 1700–1820*, Fontana, London.
7. Brewer, J., Porter, R. (eds.): 1992, *Consumption and the World of Goods*, Routledge, London.
8. Burnby, J. G. L.: 1983, *A Study of the English Apothecary from 1660 to 1760, Medical History*, Supplement no. 30.
9. Cannadine, D.: 1984, 'The Present and the Past in the English Industrial Revolution, 1880–1980', *Past and Present* 103, 131–72.
10. Cartwright, F. F.: 1967, 'The Association of Thomas Beddoes, M.D. with James Watt, F. R. S.', *Notes and Records* of the Royal Society of London XXII, pp. 131–43.
11. Cooter, R., ed.: 1988, *Studies in the History of Alternative Medicine*, Macmillan, London.
12. Cottle, J.: 1970, 1st edn, 1847, *Reminiscences of Samuel Taylor Coleridge and Robert Southey*, Highgate, Lime Tree Bower Press. Houlston & Stoneman, London.
13. Deane, S.: 1989, *The French Revolution and Enlightenment in England*

1789–1832, Harvard University Press, Cambridge, Massachusetts.
14. Eversley, D. E. C.: 1967, 'The Home Market and Economic Growth in England 1750–1800', in Jones, E. L. and Mingay, C. E. (eds.), *Land, Labour and Population in the Industrial Revolution*, pp. 206–59, Edward Arnold, London.
15. Fissell, M. E.: 1991, *The Physic of Charity: Health and Welfare in the West Country, 1690–1810*, Cambridge University Press, Cambridge.
16. Gascoigne, J.: 1989, *Cambridqe in the Age of the Enlightenment: Science, Religion and Politics from the Restoration to the French Revolution*, Cambridge University Press, Cambridge.
17. Gibbs, F. W. and Smeaton, W.: 1961, 'Thomas Beddoes at Oxford', *Ambix* IX, 417–49.
18. Glyster, G. pseud.: 1789, *A Dose for the Doctors; or the Aesculapian Labyrinth Explored*, Kearsley, London.
19. Goodwin, A.: 1979, *The Friends of Liberty: The English Democratic Movement in the Age of the French Revolution*, Hutchinson, London.
20. Gottlieb, L. S.: 1965, 'Thomas Beddoes M.D. and the Pneumatic Institution at Clifton 1798–1801', *Annals of Internal Medicine* LXIII, pp. 530–33.
21. Gregory, J.: 1770, *Observations on the Duties of a Physician*, Strahan & Cadell, London.
22. Harrison, E.: 1806, *Remarks on the Ineffective State of the Practice of Physic in Great Britain*, R. Bickerstaff, London.
23. Heberden, E.: 1989, *The Life of William Heberden*, Royal Society of Medicine, London.
24. Holloway, S. F. W.: 1964, 'Medical Education in England, 1830–1858: A Sociological Analysis', *History* 49, 299–32.
25. Holmes, R.: 1989, *Coleridge: Early Visions*, Hodder & Stoughton, London.
26. Jewson, N.: 1974, 'Medical Knowledge and the Patronage System in Eighteenth Century England', *Sociology* 8, 369–85.
27. Jewson, N.: 1976, 'The Disappearance of the Sick Man from Medical Cosmology, 1770–1870', *Sociology* 10, 225–44.
28. Jones, E. L.: 1973, 'The Fashion Manipulators: Consumer Tastes and British Industries, 1660–1800', in Cain, L. P., Uselding, P. J. (eds.), *Business Enterprise and Economic Change*, pp. 198–226, Kent State University Press, Kent State, Ohio.
29. Lane, J.: 1984, 'The Medical Practitioners of Provincial England in 1783', *Medical History* 28, 353–71.
30. Lane, J.: 1985, 'The Role of Apprenticeship in Eighteenth-Century Medical Education in England', in Bynum, F. W., Porter, R. (eds.), *William Hunter and the Eighteenth-Century Medical World*, pp. 57–104, Cambridge University Press, Cambridge.
31. Lawrence, C.: 1975, 'William Buchan: Medicine Laid Open', *Medical History* 19, 20–35.
32. Lawrence, C.: 1984, 'Medicine as Culture: Edinburgh and the Scottish Enlightenment', University of London, Ph.D. Thesis.
33. Levere, T. H.: 1977, 'Dr Thomas Beddoes and the Establishment of his Pneumatic Institution: a Tale of Three Presidents', *Notes and Records of the*

Royal Society of London XXXII, 41–49.
34. Levere, T. H.: 1981, *Poetry Realised in Nature: Samuel Taylor Coleridge and Early-Nineteenth Century Science*, Cambridge University Press, Cambridge.
35. Levere, T. H.: July 1981, 'Dr Thomas Beddoes at Oxford: Radical Politics in 1788–93, and the Fate of the Regius Chair in Chemistry', *Ambix* XVIII, 61–69.
36. Levere, T. H.: 1982, 'Thomas Beddoes, The Interaction of Pneumatic and Preventative Medicine with ISR: Chemistry' *IRS: Interdisciplinary Science Reviews* VII, 137–147.
37. Levere, T. H.: 1984, 'Dr Thomas Beddoes (1760–1808): Science and Medicine in Politics and Society', *The British Journal for the History of Science* XVII, 187–204.
38. Loudon, I. S. L.: 1983, 'The Origin of the General Practitioner' (The James Mackenzie Lecture of the Royal College of General Practitioners for 1982), *Journal of the Royal College of General Practitioners* 33, 13–18.
39. Loudon, I. S. L.: 1985, 'The Nature of Provincial Medical Practice in Eighteenth-Century England', *Medical History* 29, 1–32.
40. Loudon, I. S. L.: 1986, *Medical Care and the General Practitioner 1750–1850*, Clarendon Press, Oxford.
41. Loudon, I. S. L.: 1986, 'The Vile Race of Quacks With Which This Country is Infested', in Bynum, W. F., Porter, R. (eds.), *Medical Fringe and Medical Orthodoxy, 1750–1850*, Croom Helm, London.
42. Macfarlane, A.: 1987, *The Culture of Capitalism*, Basil Blackwell, Oxford.
43. McKendrick, N.: 1974, 'Home Demand and Economic Growth: A New View of the Role of Women and Children in the Industrial Revolution', in McKendrick, N. (ed.), *Historical Perspectives: Studies in English Thought and Society*, pp. 152–210, Europa, London.
44. McKendrick, N., Brewer, J., Plumb, J. H.: 1982, *The Birth of a Consumer Society: The Commercialization of Eighteenth-Century Enaland*, Europa, London.
45. Moravia, S.: 1980, 'The Enlightenment and the Sciences of Man', *History of Science* 18, 247–58.
46. Neve, M.: 1984, 'Natural Philosophy, Medicine and the Culture of Science in Provincial England: The Cases of Bristol, 1796–1850 and Bath, 1750–1820', University of London, Ph. D. Thesis.
47. Newman, C.: 1957, *The Evolution of Medical Education in the Nineteenth Century*, Oxford University Press, London.
48. Nutton, V. (ed.): 1989, *Medicine at the Courts of Europe, 1500–1837*, Routledge, London.
49. Parry, N. and Parry, J.: 1978, *The Rise of the Medical Profession. A Study of Collective Mobility*, Croom Helm, London.
50. Percival, T.: 1803, *Medical Ethics; or, a Code of Institutes and Precepts Adapted to the Professional Conduct of Physicians and Surgeons*, J. Johnson & R. Bickerstaff, Manchester.
51. Peterson, M. J.: 1978, *The Medical Profession in Mid-Victorian London*, University of California Press, Berkeley.

52. Pocock, J. G. A.: 1985, *Virtue, Commerce and History*, Cambridge University Press, Cambridge.
53. Porter, D., Porter, R.: 1989, *Patient's Progress. Doctors and Doctoring in Eighteenth Century England*, Polity Press, Cambridge.
54. Porter, R.: 1986, 'Medical Education in England Before the Teaching Hospital: Some Recent Revisions', in Wilkes, J. (ed.), *The Professional Teacher*, pp. 29–44, History of Education Society, London.
55. Porter, R.: 1989, *Health for Sale. Quackery in England 1650–1850*, Manchester University Press, Manchester.
56. Porter, R.: 1991, 'Reforming the Patient. Thomas Beddoes and Medical Practice', in French, R., Wear, A. (eds.), *Medicine in the Age of Reform*, Routledge, London, pp. 9–44.
57. Porter, R.: 1991, 'Commerce and Disease in Eighteenth Century England', in Halini, S. (ed.), *Commerces*, Publications de la Sorbonne, Paris, 55–73.
58. Porter, R.: 1991, 'Expressing Yourself Ill: The Language of Sickness in Georgian England', in Burke, P., Porter, R. (eds.), *Language. Self and Society: The Social History of Language*, Polity Press, Cambridge, 276–299.
59. Porter, R.: 1991, *Doctor of Society: Thomas Beddoes and the Sick Trade in Late Enlightenment England*, Routledge, London.
60. Porter, R.: 1992, 'Consumption in the Consumer Society', in Brewer, J., Porter, R. (eds.), *Consumption and the World of Goods*, Routledge, London.
61. Ramsey, M.: 1982, 'The Repression of Unauthorized Medical Practice', *Eighteenth Century Life* VII, 118–35.
62. Ramsey, M.: 1987, 'Property Rights and the Right to Health. The Regulation of Secret Remedies in France, 1789–1815', in Bynum, W. F., Porter, R. (eds.), *Medical Fringe and Medical Orthodoxy*, pp. 79–105, Croom Helm, London.
63. Ramsey, M.: 1988, *Professional and Popular Medicine in France, 1770–1830*, Cambridge University Press, New York.
64. Risse, G.: 1986, *Hospital Life in Enlightenment Scotland: Care and Teaching at the Royal Infirmary of Edinburgh*, Cambridge University Press, Cambridge.
65. Risse, G.: 1987, 'Hospital History: New Sources and Methods', in Porter, R., Wear, A. (eds.), Problems and *Methods in the History of Medicine*, pp. 175–203, Croom Helm, London.
66. Robinson, E.: 1963–1964, 'Eighteenth Century Commerce and Fashion. Matthew Boulton's Marketing Techniques', *Economic History Review* XVI, 39–60.
67. Rosenberg, C.: 1983, 'Medical Text and Medical Context; Explaining William Buchan's *Domestic Medicine*', *Bulletin of the History of Medicine* 57, 22–24.
68. Sekora, J.: 1977, *Luxury: The Concept in Western Thought, Eden to Smollett*, The Johns Hopkins University Press, Baltimore.
69. Smith, G.: 1985, 'Prescribing the Rules of Health: Self-Help and Advice in Late Eighteenth-Century England', in Porter, R. (ed.), *Patients and Practitioners: Lay Perceptions of Medicine in Pre-industrial Society*, pp. 249–82, Cambridge University Press, Cambridge and New York.
70. Smith, G.: 1985, 'Cleanliness: the Development of an Idea and Practice in Britain 1770–1850', University of London, Ph.D. Thesis.
71. Stansfield, D. A.: 1984, *Thomas Beddoes M.D. 1760–1808, Chemist, Physician, Democrat*, Reidel, Dordrecht.

72. Staum, M.: 1980, *Cabanis. Enlightenment and Medical Philosophy in the French Revolution*, Princeton University Press, Princeton, NJ.
73. Stock, J. E.: 1811, *Memoirs of the Life of Thomas Beddoes* MD, J Murray, London.
74. Thompson, E. P.: 1963, *The Making of the English Working Class*, Penguin, Harmondsworth.
75. Waddington, I.: 1975, 'The Development of Medical Ethics – A Sociological Analysis', *Medical History* 19, 36–51.
76. Waddington, I.: 1977, 'General Practitioners and Consultants in Early Nineteenth Century England: The Sociology of an Intra-Professional Conflict', in Woodward, J., Richards, D. (eds.), *Health Care and Popular Medicine in Nineteenth Century Enaland: Essays in the Social History of Medicine*, pp. 164–88, Croom Helm, London.
77. Waddington, I.: 1984, *The Medical Profession in the Industrial Revolution*, Gill & Macmillan, Dublin.
78. Wear, A.: 1989, 'Medical Practice in Late Seventeenth Century and Early Eighteenth Century England: Continuity and Union', in French, R., Wear, A. (eds.), *The Medical Revolution of the Seventeenth Century*, pp. 294–320, Cambridge University Press, Cambridge.
79. Webster, C.: 1986, 'The Medical Faculty and the Physic Garden', in *The History of the University of Oxford*, vol. v, *The Eighteenth Century*, pp. 683–724, Sutherland, L. S., Mitchell, L. G. (eds.), Oxford University Press, Oxford.

PART TWO

THE EIGHTEENTH-CENTURY PHILOSOPHICAL BACKGROUND

INTRODUCTION

Radically oversimplified, the tale we tell in these two volumes opens with an account (in the preceding section) of the moral disarray in eighteenth-century British medicine. Moral order, or, at least, a sense of workable standards of propriety, is restored at the end of the century by the almost universal acceptance of the writings of two physician-ethicists: John Gregory (1725–1773), whose influential *Lectures on the Duties and Qualifications of a Physician* was officially published in 1772; and Thomas Percival (1740–1804), whose most influential work, *Medical Ethics*, was published in 1803. The rest of the story unfolds in volume two, which relates how nineteenth-century American and British medical societies drafted codes of ethics modeled on Percival's; and how, in 1846, the first national medical society, the American Medical Association, made its first order of business the adoption of a code of ethics based on Gregory and Percival – setting a model for medical ethics which remains dominant until the mid-twentieth century.

Gregory and Percival, therefore, are the pivotal figures around whom the eighteenth-century story turns. They become pivotal because, by appealing to philosophical conceptions of virtue and moral sense, they provided answers to problems that vexed not only Beddoes, Bracken and Cleland, but most eighteenth-century practitioners – the problem of distinguishing themselves as practitioners of a "liberal profession" from mere purveyors of a trade; of distinguishing etiquette from ethics; of distinguishing scientific practitioners from quacks; and, most importantly, of finding a professional way of handling the fractious and often fratricidal disputes that threatened to destroy the lives of medical practitioners and institutions.

Anyone enamored of the myth of the Hippocratic footnote will find the

R. Baker, Dorothy Porter and Roy Porter (eds), The Codification of Medical Morality, 93–98.

concerns of eighteenth-century physicians puzzling, since the Hippocratic Oath provides a basis for resolving these issues. Yet mention of the Hippocratic corpus is conspicuously absent from the writings of Gregory and Percival. There is, to be sure, an occasional passage echoing Hippocratic ideas; thus at the beginning of his first lecture Gregory explains to his students that "[By] the *practice of medicine* ... I understand, the art of preserving health, of prolonging life, and of curing diseases." In the mind of a scholar the use of the term 'art' suggests Hippocratic influence, but Gregory himself seems unaware of this. He never makes an explicit allusion to the tradition, or mentions the name "Hippocrates" to his students. What is striking, as the authors note in the preceding section, is that no one in the eighteenth century seems to pay any heed at all to Hippocratic ethics.

If Gregory and Percival did not hark back to Hippocrates to find a medical ethic, where did they look? Only Percival is explicit about his sources of inspiration, citing a host of moral philosophers and religious moralists, and singling out for special attention two names, Reverend Thomas Gisborne (1758–1846) and John Gregory. Gregory himself is more reticent about acknowledging his sources of influence, but he makes use of the language and concepts associated with the philosophers of the Scottish Enlightenment – most notably David Hume (1711–1776) and Adam Smith (1723–1790).

More specifically, Gregory's lectures reflect the central tenet of Scottish moral sense theory, that morality is a function, not of actions and their consequences, but of motivation and character. In the first two lectures Gregory portrays the moral physician's relation to his patients as motivated by non-selfish, benevolent "sentiments," such as "humanity" and "sympathy." He characterizes "humanity" as "that sensibility of heart which makes us feel for the distresses of our fellow creatures, and which of a consequence, incites us to relieve them" ([1], p. 22). Sympathy is the sentiment which engages the humanity of the moral practitioner and makes it operational by "produc[ing] an anxious attention to a thousand little circumstances that may tend to relieve the patient" ([1], p. 22).

Although the ideal of the humanistic physician whose effectiveness derives as much from an empathic understanding of his patient (sympathy) as from medical science is commonplace today, it was alien to eighteenth-century British medicine prior to the publication of Gregory's lectures. Indeed, Gregory had to defend his conception of the humanistic physician against critics who are "callous to sentiments of humanity [and] treat this sympathy with ridicule, and represent it either as hypocrisy or as the indication of a feeble mind" ([1], p. 24). The "rough and blustering manners" affected by physicians in their interactions with patients in his day (and evident in the practitioner-patient relationship displayed in the writings and doings of Beddoes, Bracken, and Cleland) are condemned by Gregory as "generally accompany[ing] a weak understanding and a mean soul, and are indeed frequently affected by men void of magnanimity and personal courage, in order to conceal their natural defects" ([1], p. 24).

By the turn of the century Gregory's conception of the humane and sympathetic physician was accepted as the norm, as is evident from the criticisms leveled at Henry Bracken by his biographer. "In the time of the Doctor ... it was too much the custom of the Faculty, when a patient's case was critical, or become hopeless, to foretell ... how he would go on, or how and when he would die, &c. To the display of this vain, and often cruel, kind of prescience he was greatly inclined ...".[1] This biographical note was published in 1804, over three decades after the publication of Gregory's *Lectures*. Reflecting the sensibilities of medicine post-Gregory, the biographer remarks that this was no longer medical practice owing to "our progress in feeling and refinement." The standard which measures progress in practitioner-patient relationships in terms of "refinement" of "feeling" towards patients, is Gregory's; and the observation that doctors have revised their practices to conform to this standard is thus a testament to his influence.

Gregory's contemporaries were inclined to accept his views on medical ethics not merely because of the persuasiveness of his arguments, but because his conception of medical practice offered solutions to problems that vexed them, especially the eminently practical riddle of distinguishing those elements of customary medical behavior that were truly ethics (which Gregory, using the language of moral sense theory, calls "natural propriety") from those which were merely matters of etiquette and decorum. Gregory held that physicians had fundamental moral duties towards their patients. "The principal duties a physician owes his patients," he argued were grounded in the moral sentiments of humanity, patience, attention, discretion, secrecy, honor, candor, sympathy and temperance. These, Gregory claims, create "obligation[s which are] immutable, the same in all ages and nations" ([1], p. 34).

As Mary Fissell points out Gregory's somewhat abstruse view that immutable "natural proprieties" are grounded in moral sentiments provided doctors of the period with a practical test for distinguishing ethics from etiquette. Ethics rests on real moral sentiments. Etiquette, decorum, and manners are merely pretended sentiments. Thus, whereas obligations grounded in moral sentiment are real and immutable, those grounded in pretended sentiments do not create real moral duties towards patients. They are, to quote Gregory, "founded in caprice, fashion, and the customs of particular nations." So, reasons Gregory,

> There is no natural propriety in a physician's wearing one dress in preference to another ... indeed ... external formalities have been often used as snares to impose on the weakness and credulity of mankind; that, in general they have been most scrupulously adhered to by the most ignorant and forward of the profession ([1], pp. 53, 54).

Perhaps the most attractive feature of Gregory's conception of the doctor is that it offered a solution to another problem that exercised Beddoes, Bracken and their contemporaries – distinguishing medicine (as a "liberal profession") from quackery and the "sick trade." Merchants and other tradesmen engage in trade for profit, the very same motive which prompts quacks to engage in the sick trade. Gregory's humanistic physician, however, engages in medical practice from motives of humanity and sympathy, not profit. The humanistic physician, therefore, can be neither a tradesman nor a quack. Thus by making "sympathy" the operational basis of moral medicine, Gregory effectively elevates the practice of medicine to an "art," or, as Gregory tended to put the point, a "liberal profession."

Yet, Gregory reassured his students that even humanistic medicine can still promote the "private interests of its members."

> ... medicine may be considered either as an art the most beneficial and important to mankind, or as a trade by which a considerable body of men gain their subsistence .... I shall endeavour to set this matter in such a light as may shew that this system of conduct in a physician, which tends most to the advancement of his art, is such as will most effectually maintain the true dignity and honour of the profession, and even promote the private interest of ... its members ... ([1], p. 13).

Here again Gregory has recourse to the theory of moral sentiment to reconcile the practical concern of his students to earn a living with the lofty goal of serving the art – that is, of preserving health, prolonging life, and curing diseases. The humanistic practitioner "naturally engages the affections and confidence of a patient," making the patient not only more amenable to cure but also a more satisfied customer. Since sympathy is "an attention which money can never purchase" ([1], p. 22), humanistic practitioners of the art enjoy a decisive competitive advantage over their trade-minded competitors. Even the most pragmatic student can thus appreciate that virtue will find external reward.

Gregory gave his lectures to medical students, not to philosophy students. Percival wrote for physicians, not for philosophers. Thus while both writers draw freely on the language and logic of moral sense, neither explains the theory to his readers. The relationship between their work and moral sense theory (which is explored in some detail in the next section of this book) will not be evident from a direct reading of their writings. One needs, in fact, a background in the moral sense theories of the Scottish and (as it turns out) of the German Enlightenment. This section of the book consists of two chapters, one by philosopher Tom Beauchamp, the other by medical historian, Johanna Geyer Kordesch, which review, the development of moral sense theories in the Scottish and German Enlightenments.

The rationale for a chapter on British moral sense theory, especially the theories of the Scots, is evident; it provides the background which allows readers to appreciate why it would be natural for Gregory to conceptualize morality in terms of moral sentiments. Less evident, perhaps, is the reason for including a review of the development of such theories in the German-speaking world. It is somewhat insular to limit ideas to physical terrain, especially since both Gregory and Percival completed their education on the continent at Leiden (or Leyden) – which is also the *alma mater* of Benjamin Rush (1745–1813) and other eighteenth century medical ethicists. More importantly, as Kordesch explains in Chapter Five, German medical ethics of the eighteenth century was a sophisticated virtue ethic, a theory of natural propriety evinced through moral sentiments, very much like the theory that Gregory was to espouse to his students. The idea of a medical ethic grounded in moral sentiments and notions of "natural propriety" could not have been alien to anyone educated in this environment. Thus while the moral sense theory that Gregory ultimately drew on when he gave his lectures may have been Scottish, the seeds of an idea of a distinct medical ethic grounded in moral sentiments may have been transplanted from German culture.

The word 'may' is used advisedly, for we do not know which lectures Gregory (or Percival) attended in Leiden, nor do we even know what was on offer. Yet the fact of propinquity is so striking that it would be odd if Gregory did not assimilate aspects of the German-language ideal of the virtuous and sentimental physician – carrying back to Britain the germ of an idea which would sprout in the rich intellectual soil of the Scottish Enlightenment.

ROBERT BAKER

NOTE

[1] David Harley, this volume Chapter Two, p. 68.

REFERENCE

1. John Gregory, *Lectures on the Duties and Qualifications of a Physician*, London, 1771; all references are to the American edition published by M. Carey & Son, Philadelphia, 1817, p. 22.

CHAPTER 4

TOM L. BEAUCHAMP

COMMON SENSE AND VIRTUE IN THE SCOTTISH MORALISTS

Two devoted friends, David Hume and Adam Smith, have become the two most widely studied figures in the moral, political, and social climate that we now call the Scottish Enlightenment. Later generations have also regarded them as the culmination of Scottish 'moral sense theorists.' However, in even the late eighteenth century, Francis Hutcheson, Smith's teacher and colleague, and Lord Shaftesbury were acknowledged to be the founders and chief figures. The full sweep of intellectual ancestry is too vast to be catalogued here, but in addition to these four figures we need to examine a few others who played a major role during the period, even if negatively: *viz.* those of Thomas Hobbes, John Locke, and Bernard Mandeville.

The idea of a *moral sense* was a relatively neglected topic during the period, but it nonetheless played a major role in what was shared across thinkers from Shaftesbury to Smith.

I. CONSTRUCTIVE INFLUENCES ON HUME AND SMITH

In the introductory section of *An Enquiry concerning Human Understanding*, Hume provides a statement of philosophical method. This book is not on ethics, which makes the following reference to his predecessors and contemporaries striking:

> Shall we esteem it worthy the labour of a philosopher to give us a true system of the planets, and adjust the position and order of those remote bodies; while we affect to overlook those, who, with so much success, delineate the parts of the mind, in which we are so intimately concerned? ([8], p. 13).

In all later editions of this work, Hume fails to name those who so analyze the mind. However, in the first two editions he included an additional passage in which he acknowledges the influence of Francis Hutcheson, who, he says, "has taught us, by the most convincing Arguments, that

R. Baker, Dorothy Porter and Roy Porter (eds), The Codification of Medical Morality, 99–121.

Morality is ... entirely relative to the Sentiment or mental Taste of each particular being; in the same Manner as the Distinctions of sweet and bitter, hot and cold, arise from the particular Feeling of each Sense or Organ. Moral Perceptions therefore, ought not to be class'd with the Operations of the Understanding, but with the Tastes or Sentiments" ([8], p. 15). This is the foundation of the moral sense theory, and Hume is acknowledging Hutcheson as its chief exponent.

In the next paragraph in this same note, Hume expresses an indebtedness to Bishop Butler's *Sermons* for pointing to the impropriety of a simple division of the passions "into two Classes, the selfish and the benevolent" ([8], pp. 15–16). goes on in the text to acknowledge Shaftesbury, and elsewhere he identifies Locke and Mandeville for their more general contributions to "the science of man" ([1], p. xvii).

This completes Hume's picture of the major British influences on his work in ethics. It is obvious that his catalogue of influences featured the premises that we now associate with the 'common sense school.' How did it happen that these figures were the most weighty contemporary influences on the moral theory of Hume, and no less on Smith?

II. FROM HOBBES TO MANDEVILLE

A starting point in human psychology was a near certainty in the moral theory of the seventeenth and eighteenth centuries. Thomas Hobbes' theories had also come to play a centerpiece role, albeit primarily a negative one, because of his views about the individual as an actor motivated by self-interest.

(A) *"The Selfish System" of Hobbes and Locke*

Hume refers to both Hobbes and Locke as maintaining "the selfish system of morals" ([8], p. 232). Shaftesbury depicts these philosophers as explaining "all the social passions and natural affections as to denominate them of the selfish kind. Thus, civility, hospitality, humanity towards strangers or people in distress, is only a more deliberate selfishness."[1] Butler depicts Hobbes, La Rochefoucauld, and the ancient Epicureans as failing to distinguish their selfish basis for action from a more enlightened form of self-love [3]. Smith, too, gauges his opposition in terms of those "who are fond of deducing all our sentiments from certain refinements of self-love" ([22], pp. 13, 308–319).

Although these characterizations strike some contemporary philosophers and historians as shoddy caricatures of Locke and Hobbes, the latter is still frequently interpreted as egoistic and as having mistakenly presented his views as an account of moral obligation [19]. From the perspective of eighteenth-century philosophers, the most straightforward interpretation of both Hobbes and Locke was that they based social obligation – moral and political – on a foundation of self-interest, rather than duty and virtue. Hume, Shaftesbury, Hutcheson, and many of their contemporaries regarded Locke and Hobbes as having dismissed an internal psychology of natural benevolence in favor of a theory that "virtuous" conduct is motivated by self-interest.

Hobbes and Locke do not use such language to express their views on the role of self-interest,[2] but we can here circumvent this interpretative problem. Although Locke and Hobbes are giants in the history of philosophy, they are background figures for our inquiry. The first person of massive constructive influence on moral sense theory was the third earl of Shaftesbury.

(B) Lord Shaftesbury

Anthony Ashley Cooper (1671–1713), Lord Shaftesbury, was a student (but no disciple) of Locke. Shaftesbury rejected the two dominant alternatives in the moral philosophy that preceded him. The first held that morality is dependent upon religion and that morality relies on motivational premises about future rewards and punishments; the second held that morality is dependent on self-interest, which alone motivates us. As Shaftesbury saw it, fear of government and fear of God (Hobbes or theological ethics[3]) are not the only or the proper alternatives for a theory of morality. Shaftesbury reasoned that there is no morality at all if, as Hobbes has it, all acts are motivated by self-interest; for then moral motivation had been eliminated and moral qualities such as love, courage, and public interest had been reduced to their opposites. Thus Hobbes's reductionism in philosophy had eliminated morality, and Locke was no better because he "threw all virtue out of the world" with his views that fashion and custom determine the morality of cultures (except on issues of natural rights [21], II, p. 203).

Shaftesbury attempted to show that morality stands on its own, independent of these philosophers' motivational "foundations." Fear of divine judgment and personal interest seemed to him not only morally incorrect, but psychologically poor motivations. Like Kant in a later period, Shaftesbury was struck with the moral importance of purity of motive and with how central the moral evaluation of individual actions depends on the motive. A virtue needing a reward is, as he saw it, not morally worthy of a reward. Not even a saintly behavior is worthy of moral esteem if self-centered motives direct the action ([21], I, Pts. 2–4).

Shaftesbury also rejected the idea that virtue is contingent on an agreement forged through a social contract, as Hobbes had claimed. True virtue is having knowledge of and pursuing the public interest ([21], I, p. 252). The most natural of all natural principles are those that "tend toward public service and the interest of society at large" ([21], II, pp. 293–294). when the person loves virtue for its own sake, as good in itself, can the person become virtuous. This is his staple rejection of both Hobbes and the theologians ([21], I, pp. 281–282).

In what is now the most-quoted part of his philosophy, Shaftesbury grounded the naturalness-of-virtue in "the moral sense," a term he introduced into the history of philosophy, but invoked less than some of his successors ([21], I, pp. 251–254, 258–266). He does not mean that *all* persons are virtuous, but only that every creature with a "reflecting faculty" naturally has the capacity to act virtuously, and innately has a sense of right and wrong ([21], I, p. 266). He was motivated in part by his belief that Locke was short-sighted on the subject of innate ideas. But Shaftesbury did not appeal to innate ideas in the same sense Locke understood the term, as real ideas. He was defending nativism, a theory about an essence in human nature. Shaftesbury condemned any inference to the conclusion that the capacity for moral judgment is not natural ([21], II, p. 290). He also insisted that judgments of the moral sense eliminate the need for an external authority, divine or human, to make morality what it is ([21], I, p. 193, 264).

Paradoxically, Shaftesbury borrowed, in his explication of the moral sense, from Locke's empiricism and theory of reflected ideas. He argued that the moral sense delivers its conclusions not by intuition but "by reflection." Our external senses present to us "outward objects" to which we react internally through the moral sense. Kindness, gratitude, etc. result from a reflected sense that is natural, even though this sense also may be dulled (just as our external senses can be dulled) and may be perverted by custom and education: "The affections of pity, kindness, gratitude, and their contraries, being brought into the mind by reflection, become another kind of affection towards those very affections themselves, which have been already felt, and are now become the subject of a new liking or dislike" ([21], I, p. 251).

Thus, we perceive virtue and vice as directly as external objects, which is why the common person can judge immediately of the correctness or incorrectness of actions.

Shaftesbury used the moral-sense conception in an enigmatic but important analogy both to Greek ideas of balance and harmony in music and to proportion in art and architecture. Just as beauty is added by perceivers to already perceived objects, so the moral sense adds a new *perception*. He thought the mind discovers the fair and the foul much as it finds the harmonious and the dissonant, and that both truly exist, just as "musical numbers" or "sensible representations" of outward things exist.

The mind ... feels the soft and harsh, the agreeable and disagreeable, in the affections; and finds a *foul* and *fair*, a *harmonious* and a *dissonant*, as really and truly here, as in any musical numbers, or in the outward forms or representations of sensible things ([21], I, p. 251; cf. pp. 252–255 and Pt. II, sec. 2).

Shaftesbury was an esteemed intellectual in Britain and on the Continent throughout much of the eighteenth century. Montesquieu saw him as another Plato, and we have already recorded Hume's tribute. But as this century faded, so did enthusiasm for Shaftesbury's moral psychology and aesthetic morality. The start of the demise is found in his contemporary and harshest critic, philosopher-physician Bernard Mandeville, born in Rotterdam as Bernard de Mandeville.

(C) Mandeville on Turning Private Vices into Publick Benefits

Mandeville's brilliant polemical work *The Fable of the Bees: or, Private Vices, Publick Benefits*, grew and developed over more than two decades. It began as a 433-line poem entitled *The Grumbling Hive: or Knaves Turn'd Honest* (1705) [16], but came to have a loosely structured moral theory. The motivation for expansion appears to have been twofold: to fend off critics and to attack Lord Shaftesbury's views on natural virtue and self-interest as a vice. Shaftesbury's view that virtue and moral worth were eternal and immutable ([21], I, p. 255) was a thoroughly indefensible moral postulate, from Mandeville's perspective: "Two systems cannot be more opposite than his Lordship's and mine .... What Pity it is that [his noble and generous doctrines] are not true" ([15], I, p. 324).

Mandeville saw the virtues recommended by Shaftesbury as "good for nothing but to breed Drones, and might qualify a Man for the stupid Enjoyments of a Monastick Life ... but ... would never fit him for Labour and Assiduity, or stir him up to great Atchievements and perilous Undertakings" ([17], I, p. 333).

Mandeville had a shocking moral theory to offer: Vice rather than virtue maximally benefits society. He argued, in apparent alliance with Hobbes, that the motive to human action is private interest, but that the public interest is nonetheless the gainer. A "vice" is not a vice in the usual sense for Mandeville; nor is a "virtue" a virtue ([17], I, pp. 48–50). A virtue is, by definition, socially approved, "the Political Offspring which Flattery begot upon Pride" ([17], I, p. 51). Virtues are not the offspring of either reason or a noble spirit transcending a corrupt nature – the two prevalent views of the period. Mandeville could find no virtues of this description and came to the cynical thesis that it is only through praise and flattery that authorities have been able to "convince" us that public-spiritedness is noble, truly human, and reasonable, whereas private-spiritedness is selfish, low-minded, and bestial. The clever ruler uses praise and flattery, not power, to channel the citizens' energies into virtuous behavior.

A "vice," in his usage, is a material desire for luxury – a desire that he thought had built all great civilizations and had worked inexorably to the public benefit. Traits of character such as envy, a passion for luxury, and ambition are all, under this definition, vices. He thus became associated with the slogan in the subtitle he placed on his Fable: "private vices, public benefits."

Mandeville placed immediately after *The Grumbling Hive* a one-paragraph Introduction to his later and more systematic moral reflections:

> Most writers are always teaching Men what they should be, and hardly ever trouble their Heads with telling them what they really are .... I believe Man ... to be a compound of various Passions, that all of them, as they are provoked and come uppermost, govern him by turns, whether he will or not. To shew that these Qualifications, which we all pretend to be asham'd of, are the great Support of a flourishing Society, has been the Subject of the foregoing Poem ([17], I, pp. 39–40).

Mandeville is at least as concerned about moral psychology as about social reform or normative theory. Like Hobbes, Mandeville believes that humans are naturally neither sociable nor benevolent. Much of Mandeville's work is devoted to showing that various human passions such as pride are expressions of self-love and that resulting actions are to be explained in terms of our attempts to gratis self-love.

As Mandeville viewed organized society, each person is working for personal betterment and thereby promotes the welfare of all, even if inadvertently. Virtue of the sort envisaged by Lord Shaftesbury is a myth of moral philosophy. However, it was never his intention to say that unrestrained vice produces public benefit. He saw the difference between vice and crime, the latter being an extension of personal interest and vice into the domain of the individually and socially harmful. Only a certain controlled measure of vice works to the public benefit, and social managers must be sure that vice does not turn into a social evil. Mandeville wrote a book on the importance of preventing crime in order to promote the public utility, and he did not hold the thesis, sometimes foisted on him, that all vices, from theft to murder, work to the public's benefit.

It would also not be correct to interpret him as saying that vice and the consumption of luxuries should be encouraged by the state. They may at times need to be discouraged. His fundamental thesis is plainly stated in the last line of the last part of Volume I of *The Fable of the Bees*:

> [I] conclude with repeating the seeming Paradox, the Substance of which is advanced in the Title Page; that Private Vices by the dextrous Management of a skilful Politician may be turned into Publick Benefits ([17], I, p. 369).

Mandeville's attack on virtue, morality, and religion gained considerable notoriety. After 1723 his book drew responses from major figures in 18th century Britain, including William Law [13], John Dennis [5], and Archibald Campbell [4]. Berkeley attacked it in the *Alciphron* ([2], pp. 60, 73). From a moral philosopher's perspective Mandeville's work is unguarded, but it also contains many perceptive observations and is unfairly categorized as no more than an English treatise in the tradition of Machiavelli.

III. HUTCHESON AS MENTOR OF THE 'MORAL SENSE'

> The Metaphysical Philosophy of Scotland, and, indeed, the literary taste in general, which so remarkably distinguished Scotland during the last century, may be dated from the lectures of Dr. Francis Hutcheson in the University of Glasgow .... It was

from this period that Scotland, after a long slumber, began again to attract general notice in the republic of letters ([25], I, p. 428).

DUGALD STEWART

Thinkers as diverse as Hume, Alexander Carlyle, Thomas Reid, Jonathan Edwards, and Dugald Stewart regarded Francis Hutcheson (1694–1746) as the foremost moral philosopher of this period. In his first work, *An Inquiry into the Original of our Ideas of Beauty and Virtue*, Hutcheson set out to defend Shaftesbury's views against those of Mandeville. On the title page he placed his statement of purpose: "In Which The Principles of the late Earl of Shaftesbury are Explain'd and Defended, against the Author of the *Fable of the Bees: And The Ideas of Moral Good* and *Evil* are establish'd, according to the Sentiments of the Antient *Moralists*." Elsewhere Hutcheson pointedly noted that "Mr. Hobbes" and the "Christian Moralists" who appeal to "subtle selfish Motives" are his opponents, just as they had been for Shaftesbury ([11], pp. 207–209).[4]

Hume outlined the "controversy ... concerning the general foundation of Morals" that primarily attracted Hutcheson's attention:

> There has been a controversy started of late ... concerning the general foundation of MORALS; whether they be derived from REASON, or from SENTIMENT; whether we attain the knowledge of them by a chain of argument and induction, or by an immediate feeling and finer internal sense; whether, like all sound judgment of truth and falsehood, they should be the same to every rational intelligent being; or whether, like the perception of beauty and deformity, they be founded entirely on the particular fabric and constitution of the human species ([8], p. 224 & ff.).

Hutcheson picked up Shaftesbury's loosely knit ideas about a moral sense and set his philosophy in opposition to the partisans of reason: Samuel Clarke (1675–1729), William Wollaston (1660–1724), Ralph Cudworth (1617–1688), and John Balguy (1686–1748). Clarke and the friends of reason believed that moral distinctions and knowledge are conveyed by reason, giving us knowledge of the unalterable fitness of things. Unlike Mandeville, both sides in this controversy accepted moral knowledge as given, and, moreover, both believed it can be traced to a human faculty and should not be made to rest on divine revelation. The proponents in the debate also generally agreed over the role of benevolence and other moral virtues. The debate was centrally, then, over the nature of the faculty of moral knowledge.

(A) *Moral Psychology*

Hutcheson could not see how reason of itself has the capacity to arrive at moral distinctions and conclusions. If *reason* is not the source, only the *senses* remain. The moral sense is accepted in his philosophy as analogous to and actually descriptive of a sense; it is an internal as distinct from an external sense ([10], I, Sec. I). Through this sense we perceive moral qualities of action or character; just as through the external senses we perceive the sensible qualities of objects. Like vision, the moral sense is given to us all, save only for the morally blind; it is part of "our constitution and make," as Shaftesbury put it ([21], I, pp. 260–262).

Conveyed by the internal sense are ideas of reflection that arise from our perception of relations or actions. These internal ideas are either aesthetic or moral. The natural sense of beauty is constructed in Hutcheson's writings along the lines of Shaftesbury's account, and the natural moral sense is also a development of Shaftesbury's ideas. Thus, the analogy between the moral and the aesthetic was carried over by Hutcheson, but the different functions and faculties were kept distinct, allowing him to speak of two internal, superior senses ([21], I, pp. vi-ix).[5]

Through the moral sense "we perceive pleasure, in the contemplation of (good) actions in others, and are determined to love the agent (and much more do we perceive pleasure in being conscious of having done such actions ourselves) without any view of further natural advantage from them" ([10], II, pp. 101–106). Our determination to be pleased is an "amiable idea" serving both as our perception of the virtue of an action or person and also our approval. Simply put, actions are virtuous because they please us (irrespective of self-advantage). The actions or character traits that we judge virtuous are instances of benevolence, and hence Hutcheson found an indissoluble connection between the virtue of benevolence and virtue per se.

Hutcheson noted that we can judge an action virtuous even though it displeases us by negatively affecting our self-interest; and we similarly may approve a vicious act of another person. In the first case, the moral sense operates independently of approvals from self-interest. In the second case, Hutcheson encounters a more serious philosophical problem. If actions are virtuous because they please us, then why cannot any action that pleases us (even irrespective of self-advantage) be virtuous? Why are all and only benevolent actions virtuous?

By resting his moral theory on a psychological foundation, Hutcheson eliminated a possible response using either conceptual or behavioral appeals – he was unable to appeal to the concept of morality or to the wrongness of certain behaviors or actions as a determinant of morality. He was firm in the view that "the object of the moral sense is not any external motion or action, but the inward affections and dispositions which by reasoning we infer from the actions observed" ([12], BK I, Pt. I, Chap. 5).

Hutcheson compounded the problem by resting his theory of motivation to virtuous action on the same psychological foundations: We aim at virtue because virtuous acts please us; we avoid vicious acts because they displease us. This presumably accounts for our acting virtuously, as distinct from merely having moral knowledge of virtue. But the theory also entails that one who is not pleased by virtue will not have any motive to act virtuously, and of course that one who is pleased by vice will have the strongest motive to vice. Hutcheson attempted to surmount this problem by appealing to God's providential arrangement (whereby most persons are motivated to benevolence).

Hutcheson might have done better to maintain that we have a natural capacity or disposition to act benevolently and to approve and encourage benevolence, rather than to say that we perceive pleasure or that we are motivated by what pleases us. Hume would later use this argument as an alternative to Hutcheson's principles.

(B) *Personal Utility and Public Utility*

Hutcheson had more than a theory of virtue; his philosophy contained a rudimentary utilitarian moral theory of obligation. This is not surprising inasmuch as ***utilitarianism is a philosophy of controlled beneficence.*** But profound questions also follow regarding the role of reason in calculating consequences, a task for which the moral sense is not well suited. Moreover, not everyone was pleased by Hutcheson's turn to utilitarianism. The Presbyterian Church, in which Hutcheson was a probationer preacher, tried him for teaching false and dangerous doctrines contravening the Westminster Confession. The Church accused him of teaching not only utilitarianism, but the premise that moral knowledge was possible without a knowledge of God.

IV. DAVID HUME

Hume's moral theory has spawned many competing interpretations, and most of his contemporaries seem to have badly misunderstood him, some with malevolence. Hutcheson and Butler refused to defend Hume when he was under scurrilous political attack, even though they were aware of his attempt to build on their foundations. And the leading Scottish philosophers after Hume, including Thomas Reid and James Beattie, rebuked him for having denied the possibility of knowledge and value.[6]

There is something modestly bizarre about most of these interpretations of Hume. It is as if his contemporaries could never see beyond their conviction that he advanced a deadening epistemological scepticism in his early work. This interpretation led his contemporaries to *assume* that anyone so sceptical about knowledge in general must be sceptical about moral knowledge. This interpretation is particularly striking inasmuch as everyone knew Hume was opposed to the sceptical "selfish system" of Hobbes and Locke and that he adopted leading ideas in the anti-sceptical philosophies of Shaftesbury and Hutcheson.

Hume's best known work in moral philosophy is Book III of *A Treatise of Human Nature* (1739–40). Yet Hume wrote a strongly worded disavowal of this book. This statement was printed in January 1776, and then affixed as an "Advertisement" to unsold and new copies of the only philosophical writings that he had offered to the public as his authorized philosophy for thirty years, his *Essays and Treatises on Several Subjects*. In this Advertisement Hume judged the *Treatise* a seriously defective "juvenile work," at least by comparison to his later philosophy. He expressed a desire "that the following pieces may alone be regarded as containing [my] philosophical sentiments and principles."[7] [I will yield to this preference below, but will make a few references to the Treatise.]

(A) *Continuity with Hutcheson and other Predecessors*

Hutcheson and Hume corresponded over the importance of the moral virtues. In a reply, Hume noted to Hutcheson that,

> What affected me most in your Remarks is your observing, that there wants a certain Warmth in the Cause of Virtue .... I am much more ambitious of being esteem'd a Friend to Virtue, than a Writer of Taste; ... at the same time, I intend to make a new Tryal, if it be possible to make the Moralist & Metaphysician agree a little better.[8]

(B) *Sceptic, Naturalist, or Neither?*

Hume oriented much of his moral theory around virtue in the moral sense tradition. He was convinced, although his contemporaries were not, that he stood as firmly in this tradition as Hutcheson, with a strong bow in the direction of the ancients. Once certain scholars in the twentieth century realized that Hume was allied on this matter with Hutcheson, they over did the indebtedness. They claimed that Hume's philosophy in general, and his moral philosophy in particular, are little more than systematic generalizations of Hutcheson's ideas.[9]

The idea behind this interpretation is the following: Just as we are naturally constituted to sense external objects, so we are naturally constituted to feel sentiments of moral and aesthetic approval. This process does not involve rational inference. The guide to the selection of objects or ends of action are the passions. At both levels what we approve is unreflective rather than rational. So much comes from Hutcheson; Hume generalizes this limited thesis into a total philosophy by denigrating the role of reason and elevating the role of the passions. On this interpretation of Hume's work, feeling is the dominant force in even the most cognitive dimensions of human apprehension, including causal judgments in science and pure reasoning. Hume was not concerned, as a deeply sceptical philosopher would be, to show these beliefs unjustifiable or unreasonable; rather he was a psychologist seeking a *causal* explanation. The causes of all actions are the passions. Reason, being inert, does not motivate at all, not even by opposing the passions in the production of action. Custom is king, and reason is, as Hume says, the slave of the passions.

Interpreting Hume as a Hutchesonian naturalist rather than a deep sceptic may still be the mainstream interpretation. [I have elsewhere depicted this interpretation as a coarse overstatement ([1], Chap. 11), even if the premise is accepted that Hume owes more to Hutcheson than anyone else. Hume is not, in my judgment, either a moral sceptic or a mere naturalist.]

(C) *Reason and the Passions*

In his moral theory Hume then, was concerned both to refute the partisans of reason and to defend some of his predecessors' views on moral sense theory. His dismissal of reason is blunt and uncompromising:

What is honourable, what is fair, what is becoming, what is noble, what is generous, takes possession of the heart, and animates us to embrace and maintain it. What is intelligible, what is evident, what is probable, what is true, procures only the cool assent of the understanding ([8], pp. 224–226).

This passage delineates Hume's purposes and the context in which he gave a definition to the term "reason." As a result of the controversy between the defenders of reason and the defenders of the sentiments, Hume adopted a confined and cautious sense of "reason." When writing in epistemology and metaphysics, Hume tends to confine "reason" to deductive or *a priori* reason, leaving the word "imagination" for the faculty of causal or factual reasoning. Reason is thus unable to determine or discover any matter of fact. This denial of reason to inductive thought is a very narrow use. When Hume turns to practical reason and ethical theory, he modestly (and confusingly) broadens the scope of reason's powers. In ethical theory, reason is expanded to refer to both factual judgment and formal or analytical judgment. Reason is thus regarded as capable of discovering relations of ideas and matters of fact, but incapable of determining the ends of conduct.

The explanation for Hume's apparently inconsistent uses of the term "reason" rests in his anti-rationalism. He was concerned with one kind of rationalism in metaphysics and epistemology, still another kind in ethics. In the former he was concerned with rationalists such as Spinoza, who believed that causal inferences were known by reason, as a relation of entailment. In ethical theory he was concerned with rationalists such as Clarke, who believed that reason can discover eternal and immutable moral principles. Hume's strategy in both cases was to show the *impotence of reason* for these tasks. He then offered an alternative explanation that appealed, in the former case, to imagination, and, in the latter case, to the passions.

It is regrettable that Hume did not remove this (I think unwitting) inconsistency in his use of "reason," but it is also understandable once we grasp his purposes. Reason is inactive for him because it judges at most of matters of fact and relations of ideas. Value judgments, however, produce, prevent, or intentionally omit actions. Moral sentiments (that is, approbation and disapprobation), not reason, move us to action. These arguments are compelling when one considers what they were directed against. Hume did not mean that stupid moral judgments are not "unreasonable" in the broad sense that we use that term today. He often used surrogates for the term "reasonable" in this way. But the tight sense of "reason" that he employed did not allow him to say that actions deserving the term "unreasonable" are contrary to reason. No value judgment, however extreme, obscene, or brutal is against reason. It is logically impossible for these judgments to be against reason.

However, Hume did not rule out a significant role for reason in morals: "Reason must enter for a considerable share in all decisions of [moral praise and utility]; since nothing but that faculty can instruct us in the tendency of qualities and actions, and point out their beneficial consequences to society and to their possessor" ([8], p. 353). The role of reason is detector of the causes and effects of previously desired objects. Although knowledge of causes and effects cannot affect our actions unless we are already interested in them, if we are interested then causal discoveries can make a decisive difference to the actions we select or deselect. Reason therefore can modify desire and action. If one discovers by the use of reason that an outcome is not what one expected, desire may turn to an aversion. *Here reason redirects activities by changing desires.* Reason, then, is the slave of the passions in that it requires their existence; but reason is also the *informer* and *corrector* of the passions.

Hume never denies that we can have "good reasons" for our actions. We can have good reasons both in the sense of having good evidence and in the sense of acting within the bounds of the moral rules of society (rather than against or in indifference to them). To say that we ought to do what is in the common interest is a starting premise for Hume. He believes he is following his immediate predecessors and the ancients as well: What we ought to do is what morality determines we ought to do, and the rules of morality are a consensual, social matter. The idea that you ought to do something against that which is the *consensus* of a moral community is thus morally unthinkable.

The moral rules are not merely the formulation of what an individual feels. They are fixed points in a cultural matrix of guidelines and controls. Virtue, too, is socially determined, not a mere subjective standard. This account acknowledges normative statements that are correct, independent of the judgments any particular individual makes about morality.

Hume's utilitarian views, as expressed in the *Treatise*, are developed in *An Enquiry concerning the Principles of Morals* as the theory that utility alone controls what we approve or disapprove morally. This led Hume to an even more sympathetic approach to the role of reason in moral judgment, because reason must inform us of comparative utilities (in the sense of the useful consequences of pursuing one action as compared with another). In light of these facts, we choose on condition of the strongest desire. He urges us to use reason as much as possible before making a moral judgment; otherwise one is too prone to uninformed judgments ([8], Appendix I).

(D) *Virtue, Motive, and Sentiment*

Hume generally uses the (French-derived) word "sentiment" to point to certain "calm passions" that have often been associated (by other philosophers) with reason. He gives an account that follows the main lines of the moral sense school:

> The notion of morals implies some sentiment common to all mankind, which recommends the same object to general approbation, and makes every man, or most men, agree in the same opinion or decision concerning it. It also implies some sentiment, so universal and comprehensive as to extend to all mankind.
>
> When a man denominates another his enemy, his rival, his antagonist, his adversary, he is understood to speak the language of self-love, and to express sentiments, peculiar to himself, and arising from his particular circumstances and situation. But when he bestows on any man the epithets of vicious or odious or depraved, he then speaks another language, and expresses sentiments, in which, he expects, all his audience are to concur with him. He must here, therefore, depart from his private and particular situation, and must chuse a point of view, common to him with others: He must move some universal principle of the human frame, and touch a string, to which all mankind have an accord and symphony. One man's ambition is not another's ambition; nor will the same event or object satisfy both: But the humanity of one man is the humanity of every one; and the same object touches this passion in all human creatures ... ([8], pp. 339–340).

For Hume, there must be human beings with a certain subjective set of responses in order for morality to exist at all, but morality has reference not to subjective feelings but rather to universal responses as worked out in particular communities. This account permits normative statements that are correct, independent of the judgments any particular individual makes about morality.

Kant and many subsequent philosophers have been concerned that Hume's apparently *subjective* ethical theory lacks universality in its moral judgments. Hume denies this interpretation, believing that *universality* in morals is possible as a form of agreement. The capacity in human nature to reach similar reactions and opinions under the appropriate conditions is what underlies many universally shared dimensions of morality. Hume is developing both a psychological theory and a normative account about taking the moral point of view. To ascertain our obligation, he says, we naturally place our sentiments (or, in that term's broader meaning, our approvals, evaluations, recommendations, advice, and the like) into the public arena for scrutiny to see if others concur with us by forming the same reaction and adopting a common point of view. We see, he says, if we can "touch a string, to which all mankind have an accord."

When he says "morality is determined by sentiment" he means little more than that, just as colors and a sense of warmth require a certain subjective basis in order to exist, so the phenomena and the propositions of morality require certain (universally shared) responses to exist. Without those feelings, there is moral darkness, nothing at all. Hume's vision is that if reason is not the source of morals, we are left with the senses. Like vision, we have a moral sense that is given to all except the morally blind; it is part of our constitution to grasp a situation in moral terms, just as we grasp situations in visual terms. This recourse to a *moral sense* does not involve a denial of *moral knowledge*. Hume assumes that we have some form of moral knowledge traceable either to a human faculty or to a process of communal decision-making. Accordingly, Hume explains our having moral knowledge both psychologically (universal and comprehensive sentiment rather than reason is the psychological basis) and historically and sociologically (moral traditions, practices, and communal decisions account for morality).

Hume holds that we have both natural virtues (those native to the human frame, especially benevolence) and artificial virtues (those beyond human nature, but produced through a public strategy or artifice, especially justice). He offers a definition of virtue: "It is the nature, and, indeed, the definition of virtue, that it is a quality of the mind agreeable to or approved of by every one, who considers or contemplates it" ([8], p. 327). A virtue is whatever motive (the pertinent "quality of mind") that produces in observers a common sentiment of approval of that motive. Virtues are psychological traits that tend universally (because of the structure of the human constitution) to produce a pleasing sentiment of approbation in spectators. We approve acts done from motives of kindness, gratitude, friendship, parental affection, generosity, and the like because they induce our approbation. To say, then, that "X is virtuous" is to say "This quality in motivation (for example, truthfulness) stimulates approbation in (nearly) all persons."

Hume maintains that motives alone are virtuous; actions, by contrast, have no moral virtue, even if they happen to conduce to the public utility and find the approbation of spectators. For example, if a public official or head of a charitable organization provides a program of assistance to victims, the motive must be one of assistance (not one, say, of drawing a paycheck or advancing a career), if the action is to be properly classified as virtuous. Similarly, we condemn malevolent motives, not merely acts of causing harm, when making a negative moral judgment on a person's harm-causing behavior. We can speak of virtuous actions, but only derivatively for Hume: Actions are virtuous if and only if there is an underlying motive that receives universal approval. Actions are our only access to motives, but the motives underlying the actions are the true objects of our appraisal. Here is Hume's statement in the *Treatise*:

> 'Tis evident, that when we praise any actions, we regard only the motives that produced them, and consider the actions as signs or indications of certain principles in the mind and temper. The external performance has no merit. We must look within to find the moral quality. This we cannot do directly; and therefore fix our attention on actions, as on external signs. But these actions are still considered as signs; and the ultimate object of our praise and approbation is the motive, that produc'd them ([7], p. 477).

Hume's view, then, is that virtues are the motivational structures whose presence (inferred from the person's actions) produces a pleasing sentiment of approbation in an impartial spectator. To say that an action is a good action of a certain type – for example, an act of keeping a promise – means, on deeper analysis, that the person was motivated by a motive proper to the action.

Self-Love and Public Utility

In opposition to the theory of self-interest he detects in Hobbes and Locke, Hume maintains that psychological egoism is simply bad psychology.

> Tenderness to their offspring, in all sensible beings, is commonly able alone to counterbalance the strongest motives of self-love, and has no manner of dependance on that affection. What interest can a fond mother have in view, who loses her health by assiduous attendance on her sick child, and afterwards languishes and dies of grief, when freed, by its death, from the slavery of that attendance? ... These and a thousand other instances are marks of a general benevolence in human nature, where no real [self] interest binds us to the object. And how an imaginary interest, known and avowed for such, can be the origin of any passion or emotion, seems difficult to explain ([8], p. 237).

Hume reasons that social utility "recommends itself directly to our approbation and goodwill." He thinks this principle "accounts, in great part, for the origin of morality" ([8], p. 284) and that it is "the sole source of that high regard paid to justice, fidelity, honour, allegiance, and chastity: That it is inseparable from all the other social virtues, humanity, generosity, charity, affability, lenity, mercy, and moderation" ([8], pp. 295–296).

However, Hume was aware of the mixture of motives in human nature. He knew that we are far from selfless benefactors:

> There is some benevolence, however small, infused into our bosom; some spark of friendship for human kind; some particle of the dove, kneaded into our frame, along with the elements of the wolf and serpent. Let these generous sentiments be supposed ever so weak; let them be insufficient to move even a hand or finger of our body; they must still direct the determinations of our mind, and where every thing else is equal, produce a cool preference of what is useful and serviceable to mankind, above what is pernicious and dangerous ([8], p. 338).

Artificial justice is, in effect, Hume's category for rules that society constructs to thwart the elements of the wolf and serpent; natural benevolence is his category for the particle of the dove. Not surprisingly, justice and benevolence are the two most important principles in his book on the principles of ethics – his own favorite of all the books he ever published.

V. ADAM SMITH

Adam Smith (1723–1790) began his study of ethics when, at age fourteen, he went to the University of Glasgow and attended Hutcheson's lectures. He later became a student at Balliol College, Oxford, where he read Hume's recently published *Treatise*. At Oxford the book was considered atheistic and also as sceptical of all morality. Although impressed by Hume's achievement, Smith was reprimanded for reading it, and his copy was confiscated. However, this scarcely caused an irreparable delay in his intellectual development. Smith thought his teachers at Oxford so disgracefully lazy and incompetent that he was not much fazed by their anti-Humean sentiments.

Smith met Hume, probably in Edinburgh, circa 1749. They thereafter formed a close friendship that was still intact when Smith delivered his *Wealth of Nations* to Hume on his deathbed. Chronologically, Smith learned first from Hutcheson, but he admired Hume more than any previous thinker and was more indebted to Hume for his thinking on ethics, religion, and economics than to any other figure. When Hume died, Smith praised his character as follows: "His temper, indeed, seemed to be more happily balanced ... than that perhaps of any other man I have ever known .... I have always considered him ... as approaching as nearly to the idea of a perfectly wise and virtuous man, as perhaps the nature of human frailty will permit."[10]

The public was shocked at such praise of a notorious atheist and sceptic, and Smith was shocked by the public dismay: "A single, and as, I thought a very harmless Sheet of paper, which I happened to Write concerning the death of our late friend Mr Hume, brought upon me ten times more abuse than the very violent attack I had made upon the whole commercial system of Great Britain" ([23], p. 251).

Smith's constructive work in ethical theory began when he held, at age twenty-eight, a Chair of Logic at the University of Glasgow (beginning in 1751). He immediately inherited a class in moral philosophy, and in 1752 he inherited the Professorship in Moral Philosophy. The development of Smith's ideas largely occurred through a process of reacting to the theories of Hutcheson and Hume. His publication, in 1759, of *The Theory of Moral Sentiments* was the initial, but not the final, expression of this development.

Sympathy

Smith begins *A Theory of Moral Sentiments* with a chapter on sympathy. This concept shows a marked conceptual development in Smith's system. The sentiment of *sympathy* is, he believes, innate in all of us and uses another innate faculty, the imagination, to place us in the situation of being sympathetic with another. To be sympathetic is not to be motivated by sympathy, but rather to sympathize and to know that one is sympathetic.

The *moral psychology* underlying these claims involves an attack on both the Hobbesian account of human nature and certain features of the moral sense tradition. Smith regards sympathy as an underived and immediate response that cannot be attributed to self-interested motivation or self-love. But he also saw no reason to call for a special "moral sense." Moral approval can, he thinks, be reduced to the response of sympathy, and disapprobation can be reduced to, in some measure, not being sympathetic.

Smith thought we sometimes approve actions without an actual feeling of sympathy. He argued that on these occasions we cognitively apprehend a situation as one we would sympathize with, were we under a different condition than we are at present. For example, at the moment we learn a friend has lost his mother we may be in no mood or condition to feel sympathy for his plight, but we can recognize at the time that if we were not ourselves under stress or distracted by other matters, we would feel sympathy. Sympathy thus forms an important condition of action: We are motivated to give aid, comfort, gratitude and the like to others by feeling sympathetic at a socially approved level. Smith thought that our responses are also socially conditioned by the way others tutor us to have proper rather than over- or under-reactions ([22], p. 22).

Propriety

Sympathy also explains how we come to judge right and wrong actions, which Smith placed under the category of "propriety." A judgment of propriety is determined by sympathy with the motive of a person who acts. In the emphasis on motive, Smith agreed with Hume, but Smith was not as keen on the category of utility as Hume had been. Smith asked whether the utility of actions leads to our approval? He answered that it is not usefulness, but rather what is right and accurate that leads to approval. The right and accurate are qualities that we attribute to the judgment of another because it coincides with our own judgment.

Here Smith shed both the moral-sense account and the implicit utilitarianism of his predecessors. He agreed with Hume that we tend to approve as virtuous what is useful, but he denied that utility is the motive or source of the approbation. We do not, he says, praise persons for the same reason we praise a chest of drawers. The sense of propriety in approbation is deeper and more direct than the perception of *utility* ([22], Pt. IV, Sects. 1–2). His point seems to be that as a psychological fact, we do not look to utility but rather to *propriety* in making our moral judgments.

The Impartial Spectator

Smith's moral theories were expressed primarily in moral psychology rather than normative ethics. But one of Smith's most original contributions is his theory of the impartial spectator, which has important normative implications. The theory begins with an explanation of how we make first-person moral judgments. He argued that we judge our own acts and motives by placing ourselves in the position of another. I think of what another would think of me were I to perform or not perform a certain action. Moral conscience is thus shaped in an intrinsically *social* context. From childhood forward I place a value on my acts in terms of my understanding of the approval and disapproval of others. In this respect we become spectators of our own behavior, like a second self looking at the first self ([22], pp. 110–113). While we can be deceived, we are morally obligated to strive to place ourselves as objectively as possible in the light in which others view us ([22], pp. 158 ff.).

Smith generalized that we *do*, and I think he means to say *should*, strive to overcome personal interest by placing ourselves in the position of an ideal impartial spectator. It is, he maintained, the only way to correct misrepresentations of self-love. This exercise in objectivity he regarded as the proper way of forming objective rules of morality, because they are derived from the experience of many parties in society ([22], Part III).

Conclusion

The moral philosophers who we now think of as major figures in the Scottish Enlightenment each made a distinctive contribution to the history of ethics, although it has been largely camouflaged in contemporary ethical theory. Many in the twentieth century look back at these writers amazed that they minimized moral justification while maximizing psychological explanation. But these seventeenth- and eighteenth-century philosophers had a different scene of thought, a different body of assumptions, and a different set of objectives. They were confronting Scholastic Aristotelians, egoists such as Hobbes and Mandeville, natural law theorists such as Pufendorf, Pyrrhonian sceptics such as Pierre Bayle, and rationalists such as Samuel Clarke. Given this context, their arguments in defense of what we now often call "the moral point of view" are understandably presented in the vocabulary of the moral sense and the social virtues.

Department of Philosophy and
Kennedy Institute of Ethics

Georgetown University
Washington, D.C., U.S.A.

NOTES

[1] Shaftesbury's independent treatises were published in a unified collection as *Characteristics of Men, Manners, Opinions, Times* (London: 1711). The edition cited here as [21] is John M. Robertson, ed. (Indianapolis: Bobbs-Merrill, 1964 reissue). The treatises and volumes will be cited separately, and the entire volume will be cited as *Characteristics*. I have sometimes modified the text of this edition by consulting D. D. Raphael's corrections, as found in his *British Moralists* 165~1~ (Oxford: Clarendon Press, 1969).
[2] See [6], Chaps. 14–15, and *The Elements of Law*, Chap. 1; *An Essay concerning Human Understanding*, Bk. I, Chap. II, § 13, Bk. II, Chap. 28, § 5, and *Second Treatise on Government*, Ch. IX, §§ 123–24.
[3] The defenders of theology were not merely those who wrote on church doctrine and theological ethics. Shaftesbury may have considered Locke, John Tillotson, and Jeremy Taylor as defenders of the doctrine that moral motivation springs from the hope of reward and the dread of punishment.
[4] *An Essay on the Nature and Conduct of the Passions and Affections* [11]. References to Hutcheson's works are to the pages in the facsimile editions in *Collected Works* (Hildesheim: Georg Olms, 1969).
[5] *An Inquiry into the Original of our Ideas of Beauty and Virtue* [10], Preface, pp. vi–ix. In *An Essay on the Nature and Conduct of the Passions and Affections* [11], Hutcheson added to these two internal senses a public sense and the sense of honor. In *A System of Moral Philosophy* [12], he mentioned still other senses such as the sense of sympathy and the sense of decency.
[6] E. C. Mossner, *The Life of David Hume* [18], Chapter 10; David Norton, *David Hume: Common-Sense Moralist, Sceptical Metaphysician* [20], pp. 3–4.
[7] Advertisement, as printed in the 1777 edition, also printed by Strahan.
[8] *Letters of David Hume* [9], letter of 17 September 1739, to Francis Hutcheson.
[9] Norman Kemp Smith, *The Philosophy of David Hume* [24], p. vi, Chapters 1–2; Barry Stroud, *Hume* [26].
[10] *Letters of David Hume* [9], Vol. II, p. 452, Adam Smith to William Strahan, 9 November 1776 (also in Smith's *Correspondence*, ([23], p. 221).

REFERENCES

1. Beauchamp, T. and Rosenberg, A.: 1981, *Hume and the Problem of Causation*, Oxford University Press, New York.
2. Berkeley, G.: 1732, 1950, *Alciphron, or the Minute Philosopher* in *The Works of*

George Berkeley, A. A. Luce and T. E. Jessop (eds.), Thomas Nelson and Sons Ltd., London.

3. Butler, J.: 1726, *Fifteen Sermons preached at the Rolls Chapel*, London.
4. Campbell, A.: 1733, *An Enquiry into the Original of Moral Virture*, Edinburgh.
5. Dennis, J.: 1724, *Vice and Luxury, or Remarks on a Book entitled the Fable of the Bees*, London.
6. Hobbes, T.: 1651, *Leviathan, or the Matter, Form and Power of a Commonwealth*, London.
7. Hume, D.: 1740, 1978, *A Treatise of Human Nature*, (ed.), L. A. Selby-Bigge, rev. P. H. Nidditch, Clarendon Press, Oxford.
8. Hume, D.: 1748, 1772, *An Enquiry Concerning the Principles of Morals*, in *Essays and Treatises on Several Subjects*, printed by Strahan for T. Cadell, A. Kincaid & A. Donaldson, London.
9. Hume, D.: 1932, *Letters of David Hume*, J. Greig, Clarendon Press, Oxford.
10. Hutcheson, F.: 1725, 1969, *An Inquiry into the Original of Our Ideas of Beauty and Virtue*, in *Collected Works*, Georg Olms, Hildesheim.
11. Hutcheson, F.: 1728, 1969, *An Essay on the Nature and Conduct of the Passions and Affections: With Illustrations on the Moral Sense*, in *Collected Works*, Georg Olms, Hildesheim.
12. Hutcheson, F.: 1755, *A System of Moral Philosophy*, London.
13. Law, W.: 1726, *Remarks on The Fable of the Bees*, London.
14. Locke, J.: 1690, *Second Treatise of Government*.
15. Locke, J.: 1700, *An Essay Concerning Human Understanding*.
16. Mandeville, B.: 1705, *The Grumbling Hive, or Knaves Turn'd Honest*, Sam Ballard, A. Baldwin, London.
17. Mandeville, B.: 1714, 1924, *The Fable of the Bees or Private vices publick benefits*, Clarendon Press, Oxford.
18. Mosner, E.: 1980, *The Life of David Hume*, Clarendon Press, Oxford.
19. Nagel, T.: 1959, "Hobbes on Obligation", *Philosophical Review*, 68, 68–83.
20. Norton, D.: 1982, *David Hume, Common-Sense Moralist, Sceptical Metaphysician*, Princeton, Princeton University Press.
21. Shaftesbury, A., Earl of: 1711, 1964, *Characteristics of Men, Manners, Opinions, Times*, John Robertson (ed.), Bobbs- Merrill, Indianapolis.
22. Smith, A.: 1759, 1976, *The Theory of Moral Sentiments*, A. L. Macfie, D. D. Raphael (eds.), Clarendon Press, Oxford.
23. Smith, A.: 1976, *Correspondence of Adam Smith*, Clarendon Press, Oxford.
24. Smith, N.: 1941, *The Philosophy of David Hume*, Macmillan, London.
25. Stewart, D.: 1854, *The Collected Works of Dugald Stewart*, Sir W. Hamilton, (ed.), T. Constable & Co., Edinburgh.
26. Stroud, B.: 1977, *Hume*, Routlege & Kegan Paul, London.

CHAPTER 5

JOHANNA GEYER-KORDESCH

NATURAL LAW AND MEDICAL ETHICS IN THE EIGHTEENTH CENTURY

Natural law theory, easily one of the most influential ideological advances in post-Reformation Protestantism, did more than change legal and political thinking. A little discussed effect after 1690 in Prussia concerned the link between morality and professional manners in the secular meritocracy of law and medicine as these disciplines established an autonomous image. The somber picture of Lutheran and Calvinist arbiters of morality in the professions changed to the more colorful patterns of gracious worldliness, less frozen in the scrupulosity of morals and more self-assured in the temper of the reforms sought against traditionalists. The élite lawyers and doctors of the age – they were generally at universities or charged with administrative responsibilities – saw a distinct advantage in pressing for autonomy within the context of a new social style of their own. They measured autonomy by social *savoir faire*. The emphasis on the manners of the secular professions – and this is what decorum was about – suggested law and medicine were unimpeachably sovereign both in what they knew and what they were.

I will focus on the early history of this redefinition of behavior in the wake of the ideological consolidation of natural law theory where it exerted its greatest influence, at the new university of Halle in Brandenburg-Prussia between 1691 and 1747. The first date marks the appointment of Christian Thomasius to the nascent Academy (the University was inaugurated in 1694) and the latter date marks the publication of the sixth and last volume of Michael Alberti's *System of Medical Jurisprudence* (*Systema jurisprudentiae medicae*)[1]. Thomasius was easily one of the most influential law professors of the period[2], while Michael Alberti[3], Georg Ernst Stahl's successor in the chair of medicine at Halle after 1715, consolidated the link between law and medicine. This linkage established the unspoken code of professionalism for both disciplines.

The aim of the élite in both professions was to advance the image of lawyers and doctors as a professional class, setting both apart from the

R. Baker, Dorothy Porter and Roy Porter (eds), The Codification of Medical Morality, 123–139.

aristocrat and the quack[4]. While the values of the enlightened professional embraced the work ethic, his social image suggested he embraced this *comme il faut*. Thus ethical questions revert to codes of *honor* rather than codes of *conduct*.

Codes of conduct are usually spelled out and have a professional committee to watch over them, in effect a regulatory structure. Codes of honor, on the other hand, are self-regulating. They are largely internal and in the best of Enlightenment writing evidence a fine moral sensibility. In G.E. Lessing's showpiece comedy of manners, *Minna von Barnhelm* (1763)[5] about Major von Tellheim's retreat from life because of a blot upon his honor, this is much the point. Major von Tellheim had been more honorable than most in advancing cash to the defeated military and governing ranks of the Duchy of Saxony. As a Prussian he had thus spared his defeated enemy more ignominious levies. This clemency coupled with honor earned him the lively love of the aristocratic Saxon Minna von Barnhelm, who goes beyond the bounds of conduct usual for her own sex by chasing him to a Berlin inn when he is in retreat after having been discharged as a soldier and accused of having accepted bribes and lowering the spoils of war of victorious Prussia. Minna von Barnhelm, witty and intrepid, rescues the Major from his sulks, but the point made in regard to his honor is real: his own caste, military officers, seemed to suspect that he behaved other than in the manner he had assured them he had when giving his word of honor. In this suspicion alone he perceived himself rightly annihilated as an officer and not even Minna could convince him he was being oversensitive. He replies to her that he is still in possession of his reason on this point, despite the persuasive pull toward warmth and love she manages to extract from his rather colder Prussian personality.

Lessing's comedy, whose popularity in the theaters of Berlin showed how well he had touched a responsive nerve[6], illustrates what the early Enlightenment theorists thought to inculcate in terms of virtuous sensibilities. Rejecting the argument of morals as a set of strict injunctions synonymous with the Divine Will (the position of orthodox theologians), they proposed that men of reason were capable of that fine balance of sensibility and action that make up the proprieties. Once this ideal was culturally established, and Major von Tellheim represents an endearing ideal, the secular professions, the military, the law, and medicine, were expected to sustain virtue without codes or rules.

But none of this would have been acceptable for the professions if medical or legal knowledge had no authority. Such is the particular paradox initiated by the eighteenth- century Enlightenment: the avant garde lawyer and the physician repudiate the learned tradition of the university (which had invested them with authority) to become the urbane masters of their own specialties while exemplifying a strict internalized code of honor in regard to the knowledge they represented.

I. NATURAL LAW THEORY AND DECORUM

Nothing can be more practical than the art of healing, but the theoretical basis for medicine in the late seventeenth and eighteenth centuries was intellectual and a product of the learned culture of Protestantism. Philosophical ethics, moreover, was distinctly a pursuit of the educated. Radical protestant writers insisted on combining an active Christian morality with innovative ideas on belief, while orthodox Protestants stressed dogmatic purity joined to the more conservative derivation of morals from the Ten Commandments[8].

In social terms these two directions did not lead to peaceful solutions. They caused civil strife, as religious questions of community and cohesiveness tended to entangle questions of government and politics. Solutions were sought on a theoretical level, and this was one of the prime reasons for an extensive literature on natural law theory[9]. Because natural law pushed these questions into a secular paradigm, it divided theologically derived morality into two spheres, *spiritual virtues* as the domain of religion and *ethical values* as the foundation of civic life. Natural law theory of the seventeenth century systematically pursued an answer to the political quandaries of religious conflict by suggesting a secular moral framework for communal life, private morality, and just government.

Natural law theory developed in the same international context as did the tensions between radical and orthodox forms of Protestantism and in approximately the same time span[10]. The theories advanced by Thomas Hobbes in England, the Dutchman Johannes Althusius, his compatriot Hugo Grotius, and the German's Samuel Pufendorf and Christian Thomasius, to name those most influential for Prussia from the mid-seventeenth century onwards, divided political and moral considerations from one another. Althusius was among the first to separate political science from *jus naturae*, easing the way toward more utilitarian, if not

Machiavellian, concepts of government. *Jus naturae*, on the other hand, was used conceptually to systematize the basis of civic and legal order within the broader reaches of ethics. This paved the way for various interpretations of how morality would fit together with the exercise of state power, individual obligations and rights, as well as duties to the community where these went beyond those delegated to the state.

Morals, although civic in scope according to natural law theory, were not necessarily synonymous with either a legal order, the requirements of government, or the normative morality derived from traditional authority (those texts taught as traditional guides to the *philosophia practica* in the universities). The necessity for redefinitions of normative values is thus more than obvious, as is the latitude gained in moral behavior. Lessing's didactic comedy indicates one popular solution to such questions, that of moral integrity where neither law nor external regulation (such as guild rules) held sway. This is precisely the area in which the élite of the professions used the background of natural law theory to devise their own moral sensibilities.

Moral sensibilities raised the emerging secular professions to a powerful normative position within civil society. Samuel Pufendorf in his *De iure naturae et gentium* (1672) differentiated the *entia naturalia* from the *entia moralia*[11]. Nature was God's creation, while morality – as he saw it – was a separate creation inherently bound to reason and choice. Human rights were inalienable and therefore the freedom of the individual was primary to that of the state. The rights of the state (to legal sanctions or to the monopoly of force) were explicitly derived from the consensus of its citizens. Thomas Hobbes did not base his political theory on the same premises, since these were founded instead on the concept of equality in which covenants must be sought by the power of the Government[12]. The arbiter of all covenants in Hobbes' theory was the state and since he did not believe in natural goodness, the power of the state took precedence in normative disputes.

Beyond the power of the state to govern and to sanction in law, subjective freedom was not to be curtailed, thus allowing the English a much more idiosyncratic approach to behavioral norms. Christian Thomasius used Hobbes' English natural law theory to define rights and duties in regard to minimizing the requirements of state intervention, but immediately went beyond these ideas to formulate a more comprehensive system of moral expectations. His extended idea of moral behavior was meant to create a better moral order within civic society, outside of state sanction, namely to sketch in that undefined area in which men normally acted[13]. Normal activity in which men were neither criminals nor saints defined the area in which Thomasius foresaw an intriguing scope for behavioral innovations.

As Prussia rose to power in the early eighteenth century, it did this on the basis of a new class of professionals: civil servants, lawyers, doctors, university professors and writers[14]. These are the men with whom Thomasius is concerned in his writings on natural law, that theme in his writings in essence concerning itself with morality. Thomasius takes up the distinction of the Scholastics between the *forum externum*[15] and the *forum internum.* Natural law, where it deals with the outer structures of civic order, such as public acts, should adhere to prudent behavior *[Klugheitslehre]*[16]. Prudence is a political concept and Thomasius here recommends a strictly worldly wisdom, one especially useful, as his contemporaries admitted, in dealings with the Crown and at court. The internal aspects of natural justice pertain to *iustum et honestum*[17]. Both of these are more decisively moral in origin, the former reiterating proper justice in all behavior, public or private. The latter was an admonition to the kind of integrity that Major von Tellheim epitomized much later in the century.

Thomasius is holistic in his approach to all these areas of moral decorum. ('Decorum' here meaning the proper balance and evaluation of all considerations due to prudence, justice and integrity.) Natural law theory includes the more subjective and private actions available to the individual (an area Hobbes leaves open) and therefore contains within it all practical ethical matters. Thomasius also sees his interpretation of natural law theory as pertaining to the whole man, defining the natural to be both the physical and mental capacities in man. Thomasius is therefore justified, and does write on reason, will, and the affections[18]. The ethics of natural decorum, as Thomasius will predicate it for the Enlightenment, engage the whole man and therefore not only what he intends, acts upon, or thinks, but also the externals of manners and dress. The scope of his theory gives it its importance for the professions. Thomasian natural law theory is unique because it shows how to socially pattern a middle sphere of behavior that is neither censored in law nor representative of the heroic virtues (such as chastity, self-sacrifice, voluntary poverty, etc.). Moving this middle ground into the forefront of a secularly conceived civic order actually opens the way to *professional behavior.*

Natural law theory not only broke with canon law, it broke with jesuitical derivations of right behavior from the finer spectrum of whether or not one was still within the pale of morality. Thomasian developments advocated that what was right or wrong were not derivative applications of moral principles, but based, for the professional who is not a theologian or a member of the clergy, on his competence. Competence can be extended to knowing how to act with others, how to dress inoffensively, or how to cure a fever or judge a case in law. Whatever the range of competence in this freedom to move in the middle ground, it does go back to one thing: the person who chooses to express his competence has attained this and is the arbiter of it. Moral qualifications are not bounded by the intricate lineages of hierarchies of moral rules, but by a secularized principle: civic good. The secular person thus bound to his secular this-worldly community breathes a different air than pre-reformation Catholics, sectarian radicals, Calvinists, Lutherans or Puritans. Thomasius wishes to jettison traditional systemized ethics in favor of a moral integrity whose justification lies in reasoned choice. That means he is postulating that a person can evaluate and amend moral behavior, rather than examine himself as to whether he is adhering to a system of morality. Morality, in this enlightened sense, means right action based on personal insight and adequate action.

Thomasius takes up the Protestant concept of conscience, distances it from its orthodox referral to an exegesis of the Ten Commandments, and develops it in the secular sphere. The division between "matters of conscience," usually referring to religiously dictated ethics, and the "adiaphora," the middle ground of worldly actions seen as indifferent for morality, is no longer valid in the Thomasian view of decorum, which sees both in a secular and encompassing light[19].

Conscience is individualized in regard to man's capacity to reason and applied to adiaphora, those things which usually do not count, because in Thomasian thinking the whole range of secular behavior has an imminent bearing on the civic order. Not least amongst those traits, or values, are those of *autonomy*, which is here the basic value introduced, and that of *competence* (authority), which is here claimed to be the basis of individualized responsibility. Early Enlightenment Thomasian natural law theory has thus just moved the high moral ground away from institutions (Church and State) and well into the sphere of the professions and the conscientious individual. Secularized moral philosophers will argue a new systemization (Immanuel Kant, for example), but, at heart, moral education and choice had been firmly injected by Thomasius with the liberty of the moral choice in a secular community.

The thought was and is that a person of competence be allowed a judgement of competence. The middle ground at least foresees this freedom, and if Thomasius accomplished anything in his extension of natural law theory perhaps this was it. This freedom to represent competent choice was gauged to dislodge systematized ethics whose characteristics (to which Thomasius was allergic) included dogmatism, absolutism, inevitability, and infallibility.

The temper of this eighteenth century dancing in the middle ground is illustrated by one of the most extraordinary novellas of the period. Heinrich von Kleist wrote *Die Marquise von O*[20] about the rape of an officer's widowed daughter (while she was unconscious) by an opposing Russian officer whose side emerged victorious in battle. The story is not about the rape, which is never described, being only indicated by a " – ", – the most famous dash of German literary history. This most immoral and military of crimes is not the point of the narrative; virtuous rehabilitation on the other hand, is. Even before the Marquise has come to believe what to her seems impossible (since she had fainted), namely that she is pregnant, the Russian Count pleads for her hand in marriage, albeit without being able to name his crime, because the crime would preclude consent by her or by her family. So here is a story of exquisitely expressed manners, courtesy and sensibility showing not a moral fault, which rape no doubt is, but the life-changing effects of closing the wounds of honor, male and female. The Count takes it upon himself to be obstinate in reparatory virtue, despite the odds and certainly not because somebody told him to (he could have easily absconded, but instead of moralizing, a near fatal fever shows him his love). The woman in question, on the other hand, although righteously indignant, and strewing the path with obstacles, faces the life in and with her body, as well as her ability to change guilt to innocence (she accepts him), and innocence to guilt (while she was unaware of her rape, she will later accept sensual love from the same man, making her, over time, a loving "accomplice").

This complicated tale thus makes a point: no rules apply, although the moral issue is clear, when decorum acts out what those schooled in it can invent. The most telling comment on these matters appears in a twentieth-century historical novel, situated in the late Enlightenment, Christa Wolff's *Kein Ort Nirgends*[21] in which two highly cognizant and perceptive characters, male and female, namely Heinrich von Kleist and Karoline Gunderrode, meet and discover in each other the despair of sensitive moralists. At a late afternoon tea they are amongst historically identifiable persons responsible for legal codification (von Savigny), and men and women secure in the code of the social ritual. Neither Kleist nor Gunderrode, according to Wolff, whose protagonists they are, can maintain their sensibility amongst those successful moralists who represent establishment codes.

At issue here is a perception not usually associated with Thomasian natural law theory, e.g., that concurrent with his acceptance of the secularized restructuring of the civil order, his championship of Pufendorff and Hobbes, whose common aim was dovetailing individual rights with the common good, the Thomasian emphasis on decorum plainly encouraged a self-assertive individuality in moral perceptions. Lessing's and Kleist's heroines and heroes validate an inventive (and stubborn) claim to their own perception of honor, dignity, autonomy, and competence. This sort of attitude tends to rewrite codes and is notoriously difficult to systematize. It does not really fit the definition of ethics as a system of morals. Secularized morals of this sort are character-based, and in this period argued didactically on the grounds of individual reason. Von Tellheim has every reason to repudiate the mercenary ethos of spoils of war, and the spoils of war usual to the victorious (rape) are repudiated by a Russian Count intent on the only reasonable antidote to male dishonor, *reparatory virtue*.

II. MEDICINE AND DECORUM

Do the finer points of secular virtuous sensitivity pertain to medicine? One can argue this case within the same boundaries as those pertinent to the social intent of Thomasian natural law theory. First of all, if it is accepted that Thomasian ideas went beyond ethical regulation and introduced elements of individual responsibility and moral competence, his medical colleagues in Halle were not only aware of it, they actively engaged in its propagation.

In 1694, when the medical faculty was established along with the University, it was both rebellious and anti-traditional[22]. Both professors of medicine appointed were innovators. Innovation, in this context, does not mean technically useful discovery, but the introduction of new methods in a nascent science, medicine. Friedrich Hoffmann (1660–1742) will insist on exploring and teaching medical research and practice on the basis of investigating material cause and effect relationships (matter theory, physiology, and chemistry) integral to the mechanical philosophy[23].

Georg Ernst Stahl (1659–1734) was already suspect as an anti-traditional innovator when he was appointed[24]. In both chemistry and physiology he made people nervous. He broke with the alchemical tradition in chemistry and he broke with mechanist-somatic physiology despite the Dutch and the philosophers (Newton, Descartes, Leibniz). Immatriculation in medicine in Halle meant that deregulation from authority and codes was an acceptable part of education. If you were a follower of Stahl you were trained to be unhappy with the methods of derivative proof and systematization. If you were advancing medical knowledge with Friedrich Hoffmann you were going to see medicine secularized in no uncertain terms. In his *Medicus politicus*[25] he advocates the persuasive techniques of the courtier, the urbanity, the manners, the dissimulation, the authority of Thomasian decorum with a touch more of the persuasive and a touch less of von Tellheim's moral sensitivity. None the less: this is the middle-ground of self-assertive decorum. It is a mode of conducting oneself (effectively) with no recourse to a systematized ethics. Political it is, as was natural law theory, in wanting to reconstruct civil society. The widening of this kind of political and cultural know-how in the medical field then belongs to books like the *Galante Patiente*[26] by Johann Daniel Longolius (a student of Stahl). The mannerly physician is a touch above the client and well away from the quack.

The inwardly sensitive variant of a medical ethos, like the literary creations, belong to the more radical followers of Stahl, like Johann Samuel Carl, who wrote the *Decorum Medicum*[27], a compilation of attacks on medical insufficiencies, lack of care for the poor, lack of hospital care, lack of sufficient knowledge, high-cost remedies and money-grabbing. The high moral tone is here argued, not stipulated or codified. If Carl had been a playwright or a novelist of the Enlightenment, we would have an anti-Molière bourgeois tragedy of a scrupulous physician's moral choices.

Obviously the consensus in morals here evident is an innovative piece of professionalism not happy with contemporary practice nor with unexamined traditions, of the kind which circumvent moral considerations, or even worldly ones, in medicine. Still, they were not relying on the Hippocratic Code or a College of Physician's rules of conduct[28]. The reason for this is the fundamental assumption that the quality of medical knowledge elects one to be competent in moral professional choice. Reason, of the 18th- century, anti-authoritarian kind, was their favored *a priori* legitimization for the competence of new knowledge as well as acceptable morals. If the capacity to reason is no longer integral to human action or moral definition, a coercion is necessary: codes.

The specific co-operation with Thomasian desiderata in the field of law became extant with Stahl's successor in Halle, Michael Alberti (1682–1757). Michael Alberti, more than anyone else, succeeded in merging the concerns of useful knowledge, professional authority, and decorum, as Thomasius had introduced them. Alberti's persistent and systematic exploration of the interface between law and medicine began in the second decade of the century. Most representative of his work in this direction was his *System of Medical Jurisprudence*, a work of six volumes (published in successive volumes from 1725 to 1747)[29]. These comprise thousands of pages in three major areas: theoretical summaries and statements; a selective and sometimes annotated review of the literature; the consilia or case studies involving medical evidence crucial to court trials. All volumes are well indexed, as their primary value was exemplary and consultative. The large number of consilia (advice on court cases) dovetail with extensive juridical publications of the same kind, a genre in which the Halle law faculty produced some 4000 model cases for study between 1693 and 1749 in published works alone[30].

In addition Alberti wrote the *Commentatio in Constitutionem Criminalem Carolinam Medica*[31] in 1739 in which he treated extensively both legal-medical questions in German law as well as separately writing on particular matters specific to medical testimony. Many of his expositions link up with legal changes discussed by Thomasius, such as a re-evaluation of torture as an inquisitional method[32]. Medical testimony in legal cases was frequent since paternity, infanticide, inheritance, annulment of marriage, murder, poisoning, and assault turned on medical evidence.

Medical literature specializing on juridical matters was frequent enough before 1725. In fact, Alberti's and Thomasius' writings are suffused with it. Their approach differs, however, from previous citations of precedent because it is severely critical and not given to a continuation of traditional authority[33]. Again, this established their own professional competence. The post 1700 battle for worldly conduit for university educated professionals was carried over into medicine in particular, through Alberti. If Thomasius wrote overtly on *decorum*, *honestum*, and *justum*, Alberti practiced it and admonished his medical colleagues to do the same.

In 1725, Alberti begins publication of his six-volume *System of Medical Jurisprudence*. In the introduction to these volumes, one can follow his close ideological partnership with Thomasian natural law reform. In the same manner as his colleague in the law faculty, Alberti reviews the older authoritative literature on his subject[34]. He submits it to the same eclectic and critical sifting used by Thomasius to reject what is no longer in line with the empirical advance of his own and the Halle medical faculty's modern accomplishments in medical reform. Alberti's heavy reliance on case study material is again not traditional[35]. It reflects the new epistemological method, also used by Thomasius in reforming law, in which the emphasis on case studies forms the basis for legal knowledge. The correct assessment of the case is the epistemological foundation for precedent law, an innovation the eighteenth-century advocates of natural law theory wished to introduce against the Continental codes embodied in Caroline and Roman law[36]. The emphasis on the case study in medical testimony served much the same purpose.

It established modern medical knowledge on the basis of empirical fact and with it the incontrovertible authority of the professional. He was now in a new position to make responsible medical decisions on his own ground. Vested with this authority there need be no appeal either to the canon of university-taught medicine or to theologically-founded morality. Ethics could comply with either one, but it could also argue another case. *Lehrfreiheit*[37] had finally established what was not self-evident in 1690, namely that professional knowledge had its own inherent right to formulate ethical judgements. In a word: medical ethics was about to be established, fathered by natural law theory.

There was, therefore, something like a common cause in the university. This readiness for innovation, coupled with phobia for authority, supported one of the most liberal of university statutes imaginable at Protestant universities. The law faculty drafted general university statutes remarkable for their exclusion of an oath to uphold "right-thinking" or *recta ratio* (the principles of the Protestant faith). According to these statutes it was also permissible to teach on subjects outside one's own faculty, the first instance of *Lehrfreiheit* in Germany, and this engendered a freedom from the censorship otherwise practiced in universities. The academic freedom possible in Halle prepared the ground for the ideological changes engendered across faculty lines.

When the sixth volume of Alberti's *System of Medical Jurisprudence* appeared in 1747, he was hammering home the professional conduct implied by *Lehrfreiheit.* Professional conduct and professional internal morality required the following attention to principles: (1) to judge medical facts presented by a case on an empirical basis, not adding any hypothesis; (2) to admit to ambivalence and medical doubt; (3) to summarize all the evidence succinctly and to show how it leads to a medical judgement[38]. Professional conduct is strictly in line with an empiricist's truthfulness. "*Hic non iocandi locus et tempus est*", as Alberti writes[39]. Through this strict behavioral admonition Alberti publishes, in effect, his advice on medical decorum. No gentleman physician behaves with the flourish of the quack. He preserves his dignity intact where he cannot judge the case. He thereby acts as a creditable expert witness. The narrow road to virtue is here indicated and recommended. It undermines exactly that shade of disbelief that medicine suffered in a culture open to empirics, and specifically to a public mockery of medical failure through the disestablishment of medical certainty.

Alberti knew he was well advised to do this. The role of the physician as expert witness in judicial cases was under severe scrutiny of council for the Defense[40]. Alberti's prefaces make clear how much he abhorred the Defensor, since it was this man's office to throw medical evidence in doubt. The strategy of the Defense against convictions in cases of infanticide, murder, assault, poisoning, etc. – those cases which turned on medical evidence – was simply to demolish medical testimony and certainty. The Defense seems to have had a field day with the older medical literature on cases, since selective citing of these could establish almost any opinion as contradictory[41]. Hence medical honor, authority and competence were at stake. Alberti's harping upon checking and rechecking medical evidence thus has crucial significance. The particular doctor may lose his private reputation as a physician when his cures go badly, and he may lose his income, but in a public testimonial act central to law and justice, the very core of medical authority is touched upon.

Alberti's prefaces and his selection of model cases in the *System of Medical Jurisprudence* should be evaluated in this light. They are a showcase for the integrity of the medical profession. Furthermore, in almost all of the cases the medical faculty has to revise or substantiate the prior findings of medical colleagues. In infanticide cases, for example, the fact of live birth and the probable cause of death are first established by the city physician and a surgeon[42]. Their report, along with the other documents in the case, were sent to the faculty to assess. The faculty testimony then had the delicate task of presenting its own conclusion, often having to point out what was not observed properly beforehand[43]. Council for the Defense could always exploit the loopholes. No codes regulated the attitudes of medical professor to city physician and surgeon in these public acts. It is a measure of Alberti's establishment of medical decorum that he sets standards, namely those of reticence, explicitly factual argument, remonstrance not with colleagues but with their findings, and the insistence on the highest quality of medical knowledge available.

The ethics of professional responsibility and an appeal to values that represent a calling rather than a career are finely elucidated in Alberti's teaching and programmatic writing. It is perhaps not amiss to stress that at a reform university such as Halle, pride of place was going to be given to high standards of moral and social conduct and not to the trappings of success. Even Thomasius' urbane challenge to the habitual well-heeled stuffiness of academia was not a problem of style. The worldliness of the new professionalism clothed very serious men with no other means of making their case than to show that their inward moral habits were urbane enough to challenge the sense of failure wished upon them by those that opposed them, those who had tradition or the play of the market at their back. A case in point are the financiers for Frederick the Great's military campaigns, who lived in the grand style while they advanced credit and fell into not so grand disgrace with their own bankruptcies[44] – there style plastered over the ethic of ruthless advantage. The point of professional urbanity lies in the adherence to the principles of civic virtue, as defined by natural law theory, where the professions are ethically bound to advance communal good rather than private fortune. Doctors and lawyers were not primarily money makers. Secularized natural law retained the value of responsibility in office as an ethical good[45]. One should recall that most of Thomasius' writings and all of Alberti's prefaces were in polemicist form, directed against those who neither liked the nature of reason (natural law theory) nor its pietist version of active Christian support of communal good.

Natural law theory created the co-operation between law and medicine in the public visibility of court trials. But it also advanced a social pose not bound entirely to the matter in hand. Because the representatives of natural law theory had embraced habits and a turn of mind that joined empiricist usefulness with gentlemanly behavior (the middle ground of

Thomasian decorum) they embodied a professionalism imbued with more scope than technical proficiency. In the public arena of court trials (and the published literature attached to these dealings), theory was shown to be very practical indeed. It would be useless to find a code for an inward attitude which clothes the mind as well as the body with an eighteenth century desire for gentlemanly propriety. Regulations and codes are produced in more decadent phases precisely when an inner consensus no longer holds. At least for three quarters of a century, morals and manners were as much a part of professionalism as they were the topic of fictional representation.

What is more, to return to an earlier point, manners were creative morals, as the finest of manners are not those of the dinner table or books of social etiquette, but the kind suffused with decorum. The person, not the logic or the code, chooses to act. Enlightened moral freedom, as evidenced in the literature and the people mentioned throughout this essay at the very least ensured that people knew how they acted, could explain it, and stood up to it.

Wellcome Unit for the History of Medicine
Glasgow University
Scotland

NOTES

[1] Alberti, Michael, *Systema jurisprudentiae medicae, quo casus forenses, a JCtis et medicis decidendi, explicantur omniumque facultatum sententiis confirmatur, in partem dogmaticam et practicam partitum, casibus relationibus, judiciis, responsis et defensionibus juridicis et medicis forensibus specialibus illustratum.* Halae, 1725 (Vol I); Vol II, 1729; Vol III, 1733; Vol IV, 1737; Vol V, 1740; Vol IV, 1747.

[2] Assessments of his influence in: Notker Hammerstein, *Jus und Historie*, Göttingen, 1972; Michael Stolleis (ed): *Staatsdenker im 17 und 18 Jht*, Frankfurt am Main, 1977; Ernst Bloch, *Natural Law and Human Dignity*, London 1986, Rolf Lieberwirth, *Christian Thomasius: Sein wissenschaftliches Lebenswerk, Eine Bibliographie*, Weimar, 1955.

[3] No good current monograph on Alberti exists; a factual account of his life and works in: W Kaiser and Arina Volker Michael Alberti (1682–1757), *Wissentschaffliche Beitrage der Martin-Luther Universität Halle-Wittenberg*, Halle, 1982.

[4] On the entrepreneurial ethos of the quack see: R Porter, *Health for Sale, Quackery in England 1660–1850*, Manchester, 1989.

[5] A good edition with a fine introduction on the social meaning of money and military practices in Prussia: Joachim Dyck, *Minna von Barnhelm oder: Die Kosten des Glucks*, Berlin, 1981.

[6] Ibid, p. 213–220

[7] Geyer-Kordesch, J, "Georg Ernst Stahl's radical Pietist medicine and its influence on the German Enlightenment" in: A Cunningham and R French (eds) *The Medical Enlightenment of the Eighteenth Century*, Cambridge, 1990. In a different vein: John Henry "The matter of souls: medical theory and theology in seventeenth-century England", in: R French and A Wear, *The Medical Revolution of the Seventeenth Century*, Cambridge, 1989.

[8] The debate on *philosophia moralis* and *theologia moralis* (revelation as the source of authority) is briefly discussed in relation to the work of J F Buddeus in: Timothy John Hochstrasser "Natural Law, its Historiography and Development in the French and German Enlightenment circa 1670–1780" (Ph.D. diss., Downing College, Cambridge, 1990) p. 232. Buddeus, however, is close to Thomasius while the orthodox position refers even more strongly to the authority of the Bible.

[9] This is substantiated in particular when one looks at the biographies of the proponents of natural law in connection with their works. See: Michael Stolleis, *Staatsdenker des 17. und 18. Jhts*; Wolfgang Rod, *Geometrischer Geist und Naturrecht*, München, 1970: Wilhelm Schmidt-Biggemann, *Topica Universalis, Eine Modellgeschichte Humanistischer und Barocker Wissenschaft*, Hamburg, 1983.

[10] Ilting, Karl-Heinz, "Naturrecht", in: O Brunner, W Conze, R Koselleck, *Geschichtliche Grundbergriffe, Historisches Lexikon zur politisch-sozialen Sprache in Deutschland*, Bd. 4, Stuttgart, 1978, pp. 245–313. Some of the thoughts in the following paragraphs are from Ilting's article on natural law. However, the interpretation of Thomasius' relation to Hobbes is my own.

[11] Ibid p. 291 ff.

[12] Ibid p. 278 ff.

[13] Thomasius, Christian, *Einleitung zur Vernunfft=Lehre*, Halle, 1691; *Ausubung der Vernunfft=Lehre*, Halle, 1691; *Enleitung zur Sittenlehre*, Halle, 1692; *Ausubung des Sittenlehre*, Halle, 1696.

[14] Rosenberg, Hans, *Bureaucracy, Aristocracy and Autocracy. The Prussian Experience 1660–1815*, Boston, 1966; Rudolf Vierhaus, *Deutschland im 18 Jht.: politische Verfassung, soziales Gefüge, geistige Bewegungen*, Göttingen, 1987; R Vierhaus, *Deutschland in Zeitalter des Absolutismus 1648–1763*, Göttingen, 1978.

[15] Hammerstein, Notker, *Jus und Historie*, Göttingen, 1972, p. 73 ff.

[16] Thomasius, Christian, *Kurzer Entwurf der Politischen Klugheit, Sich Selbst und Anderen in allen Menschlichen Gesellschaften wohl zu rathen und zu einer Gescheiten Conduite zu gelangen*, Leipzig, 1710 (reprint 1971); *Einleitung Zur Hoff = Philosophie, Oder, Kurzer Entwurff und die ersten Linien von der Klugheit zu Bedencken und vernunfftig zu schliessen ...*, Berlin, 1712.

[17] These concepts best developed in: Christian Thomasius, *Fundamenta Juris Naturae et Gentium*, Halle, 1705 and *Grund-Lehren des Natur- und Volcker-Rechts .... In welchem allenthalben Unterschieden werden die Ehrlichkeit, Gerechtigkeit und Anstandiqkeit*. (The 1709 German translation of the *Fundamenta*.)

[18] If one takes the writings on natural law, on urbane manners (*Klugheit*), on morality (*Sittenlehre*), and on the soul as the major preoccupations of Thomasius until the end of the first decade of the 18th century one sees him as more than just an advocate of juridical natural law theory. Historians tend to fragment his thinking rather than

seeing his work as an argument for a new social order.

[19] The basic achievement of the *Fundamenta Juris Naturae et Gentuim* is to deny the Bible legal authority and thereby split canonical from secular law.

[20] von Kleist, Heinrich, "Die Marquise von O", in: Erzählungen, München, 1968, pp. 94–130.

[21] Wolff, Christa, *Kein Ort Nirgends*, Darmstadt, 1979.

[22] Geyer-Kordesch, J, "Georg Ernst Stahl's radical Pietist medicine and its influence on the German Enlightenment", in: A Cunningham and R French, *The Medical Enlightenment of the Eighteenth Century*, Cambridge, 1990, pp. 67–87.

[23] On Hoffmann, see: R French, "Sickness and the soul: Stahl, Hoffmann and Sauvages on Pathology", in: *The Medical Enlightenment* (as above), pp. 88–110.

[24] Geyer-Kordesch, J, Georg Ernst Stahl: *Pietismus, Medizin und Aufklärung im 18 Jahrhundert in Brandenburg-Preussen*, 1988 (MS).

[25] Hoffmann, Friedrich, *Politischer Medicus oder Klugheits=Regeln*, Leipzig; German translation of *Medicus politicus sive regulae prudentiae secundam quas Medicus juvenis studia sua et vitae rationem dirigere debet ...*, Leiden, 1738.

[26] Longolius, Johann Daniel, *Galanter Patiente, oder: Philosophischer Unterricht, wie sich ein Krancker so wohl gegen sich selbst, als gegen andere, nett und galant auffuhren soll*, Büdissin, 1727.

[27] Carl, Johann Samuel, *Decorum Medici*, Büdingen, 1719.

[28] Regulation through royal decree came in 1725 (see: J Geyer-Kordesch, "Court Physicians and State Regulation in Eighteenth-century Prussia: The emergence of medical science and the demystification of the body", in: Vivian Nutton (ed) *Medicine at the Courts of Europe 1500–1837*, London, 1990 pp. 155–181), but this concerned qualifications rather than decorum or the Hippocratic Oath. Hippocratic writings were thought of highly during this time of reform, the Oath, however, does not figure prominantly in these.

[29] See footnote 10.

[30] Schubart-Fikentscher, Gertrud, *Hallesche Spruchpraxis, Consiliensammlungen Hallescher Gelehrter aus dem Anfang des 18 Jahrhunderts*, Weimar, 1960, p. 1.

[31] Alberti, Michael, *Commentatio in Constitutionem Criminalem Carolinam Medica*, Halae, 1739.

[32] Ibid: several specialised sections.

[33] Alberti, *Systema jurisprudentiae medicae*, Tomus I: Preface by Christian Thomasius also passim and in the prefaces to the volumes written by Alberti.

[34] Ibid, Pars I: gives the main general introduction to forensic medicine and its literature; Pars II goes into the cases.

[35] See: my forthcoming article on the epistomological and political meaning of empirical case studies "Medizinische Fallbeschreibungen und ihre Bedeutung im Fruhen 18. Jht" to be published in: *Medizin, Geschichte und Gesellschaft*, Stuttgart.

[36] Hammerstein, Notkar, *Jus und Historie*, Göttingen, 1973 p. 173 ff.

[37] Ibid. p. 167.

[38] Alberti, *Systema jurisprudentiae medicae*, 1725–1747: summarized from the prefaces to each volume.

[39] Ibid, Vol 5 (1740), preface (unpaginated).

[40] Ibid, Vol 4 (1737), preface (unpaginated).

[41] Ibid, but this is a continual theme throughout.

[42] Ibid, for example, the infanticide case in Vol 6 (1747) of 26 February, 1743.

[43] Ibid, but again a theme throughout the case material.

[44] A very good exposition of this in: Joachim Dyck, *Minna von Barnhelm oder: Die Kosten des Glucks*, Berlin 1981, the chapter entitled "'Einige Kapitale werden jetzt mitschwinden': Geldgeschafte", p. 59–68.

[45] "De officiis", "on duties" are prominent chapters in Thomasius' *Fundamenta Juris Naturae et Gentium*, Halle, 1705; Duties in Prussia were linked with office, and therefore also with official prestige, of the universities or the state, and were not primarily linked to behavioural advantage in free trade among entrepreneurical lawyers or doctors. The honour of the military officer can thus be seen to be closely linked to the honour of professionals in office since they were all representative of the state.

PART THREE

THE FORMALIZATION OF MEDICAL ETHICS

INTRODUCTION

The background to the eighteenth-century formalization of English-language medical ethics was the malaise of medical practitioners (explored in Part One) and Enlightenment theories of moral sense and sensibility (reviewed in Part Two). The formalization itself, however, was primarily the work of the two individuals who are the focus of this section, John Gregory and Thomas Percival.

It is important to appreciate that the project attributed to Gregory and Percival is an artifact of retrospection; they did not originally see themselves as engaged in anything as lofty. Gregory's *Lectures On the Duties and Qualifications of Physicians* was originally lectures to Edinburgh medical students, as the title indicates. In his "Advertisement" Gregory claims the lectures were only published to correct printed transcripts that had been put into circulation by his "pupils" and which exhibited his words "in negligent dress." His only stated ambition for the new edition is to hope that it would prove "useful" to students and "the younger part of the Faculty," and promote "the true interests of Physic."

Percival had, in the end, more lofty ambitions. As John Pickstone argues in Chapter Seven, however, the character of *Medical Ethics* was set in its earlier incarnation of 1792, when it was drafted as a set of rules to settle a dispute that was disrupting the Manchester Infirmary. In 1794 Percival reformulated and expanded these parochial rules, recasting them as a model for other hospitals. These were printed and circulated privately under a title – "Medical Jurisprudence." Nine years later, in 1803, he again revised and expanded the rules, publishing them under the different title of "Medical Ethics." Percival characterizes *Medical Ethics* as a code of governance for the faculty of hospitals, designed to insure "that the official conduct and mutual intercourse of the faculty might be regulated

R. Baker, Dorothy Porter and Roy Porter (eds), The Codification of Medical Morality, 141–144.

by precise and acknowledged principles of urbanity and rectitude." Almost from the moment of publication, however, segments of Gregory's *Lectures* and of Percival's *Code* were extracted from their original context and appropriated for other purposes. Percival's treatment of Gregory's admonitions to his students is illustrative. In Chapter Seven, Laurence McCullough catalogues the virtues Gregory admonishes his students to cultivate as physicians: among them are attention, fidelity, humanity, secrecy and steadiness. Article I of Percival's code of "professional conduct in private or general practice" lists these same virtues in a "moral rule of conduct" stipulating that "Every case, committed to the charge of a physician or surgeon, should be treated with attention, steadiness and humanity ... secrecy and delicacy ... fidelity and honor." What was originally an exhortation is thus transformed into a rule of conduct.

Percival's code suffered a similar fate, when it, in turn, was removed from the hospital setting and appropriated as a code for medical societies. Especially in the United States, new societies were being formed at the municipal, county and, eventually, the national level and they typically prefaced their charter with a code of ethics written in language drawn from Gregory and Percival. Transplanting Percival's code to American soil, however, required radical reconceptualization because the British medical division of labor into physicians, surgeons, and apothecaries presupposed by Percival was non-existent in the new world. Consequently, although physicians treated both Percival and Gregory as sources for their codes, few read his code in its entirety – why would anyone in a country which had no apothecaries read a code dealing with the physician-apothecary relationship?

Throughout much of the nineteenth and the early part of the twentieth century, therefore, Gregory and Percival were treated as revered founders of medical ethics, classic authors whose achievements were widely known, whose works were often cited, but whose writings were seldom actually read (and certainly not read in their entirety). The subsequent hagiographic treatment of Gregory and Percival set the stage for a critical, revisionist reappraisal (the historiography of which is developed in some detail in Chapter Eight). What is common to the revisionist readings of Jeffery Berlant, Ivan Waddington, and other scholars is the attempt to deflate the saintly image of Gregory and Percival by taking these worthies off the lofty philosophical plane and placing them in a concrete socio-economic context. Reexamining their writings in this light, the revisionists recast Percival in particular as a conservative defender of the *status quo*. Some even charged that his so-called "ethics" was nothing more than an etiquette developed to settle squabbles between practitioners and to preserve the medical division of labor.

The three essays which form Part Three can be considered "post-revisionist" readings of Gregory and Percival. In their various ways, the authors seek to reappraise received revisionist interpretations. In Chapter Six, for example, philosopher Laurence McCullough reasserts the propriety of treating Gregory as an intellectual conversant with, and actively engaged in, philosophical reflection. McCullough establishes that Gregory's use of the term 'sympathy' fits precisely the pattern of usage in the writings of Gregory's contemporary, David Hume, but is notably different from that of another contemporary moral sense theorist, Adam Smith.

In Chapter Seven, historian John Pickstone addresses the revisionists' reading of Percival. Pickstone fully accepts the project of placing Gregory and Percival in a socio-economic time and place. He argues, however, that Berlant and Waddington err in taking Britain in the Age of Reform to be the appropriate time and place for Percival. The relevant place is quite particular, it is Manchester, not London. The relevant time is equally specific, the years between 1792 and 1803. In 1792, at the request of the trustees of the Manchester Infirmary, Percival developed the rules of governance that would ultimately be published as *Medical Ethics* in 1803. After reviewing events in Manchester and at the Infirmary in that decade, Pickstone asks a trenchant question: If, as Percival states, *Medical Ethics* was essentially complete in 1795, why did he wait until 1803 to publish it? Because, replies Pickstone, there was something Percival wanted to communicate at *that* time. This was a period in Manchester's history when Enlightenment projects, like the Infirmary, were under siege. Publishing *Medical Ethics* was thus a means of reasserting Enlightenment ideals at a time and in a place which had become profoundly hostile to them. Ironically, Pickstone grants the revisionist thesis that Percival's intent was conservative – to defend and stablilize the medical establishment. But, he points out, to defend established medical institutions in the *Manchester* of *1803*, was also to defend the ideals of the Enlightement – which helps to explain Percival's appeal to later generations, especially in the United States (a nation founded on Enlightenment ideals).

The American appropriation of Percival's code is a central focus of Chapter Eight in which philosopher Robert Baker directly challenges the revisionists' reading of the American Medical Association's 1847 code of ethics and its intellectual progenitor, Percival's ***Medical Ethics***, as mere professional etiquettes. Baker contends that although the revisionists' criteria for "genuine ethics" are overly stringent, if Percival's code is read as an eighteenth-century document, it actually satisfies them. On Baker's reading, Percival is a syncretic writer whose code is an amalgam that consists, not only of Enlightenment ideals of egalitarianism and humanism, but also of precepts developed by hospital medical practitioners to facilitate professional collaboration and dispute resolution. By incorporating these practical precepts into his code, Percival was forced to state, in a concise manner and at a relatively abstract level, an intersubjective morality, radically different from anything found in Gregory's writings (or, for that matter, in the works of the Hippocratics). Percival's ethics, therefore, is a revolutionary landmark, indicating the transformation of medical ethics from a character-based, subjective ethics, to intersubjective morality of peer review, of hospital rounds, and of a collaboration, which is standardly practiced in contemporary teaching hospitals.

ROBERT BAKER

CHAPTER 6

LAURENCE B. MCCULLOUGH

JOHN GREGORY'S MEDICAL ETHICS AND HUMEAN SYMPATHY

John Gregory, of Aberdeen and Edinburgh, without doubt stands as one of the two major figures of British medical ethics in the latter half of the eighteenth century – the other being Thomas Percival of Manchester. Gregory's distinctive contribution is that, by steeping himself in the moral philosophy of the Scottish Enlightenment, he provides a plainly philosophical basis for medical ethics. Living at a time when the age of manners had all but ended, Gregory could not rely on received views about good conduct and the right order of society [6]. Only more a reliable foundation would do. By setting about to provide such a foundation in terms of the moral philosophy of the Scottish Enlightenment, Gregory marks himself – from our perspective – as the first modern figure in Anglo-American medical ethics.

As can be seen from the text (see section III below), Gregory bases his medical ethics on the concept of *sympathy*, a central concept of the moral sense theories of the Scottish Enlightenment. The two main proponents of this concept were David Hume [19] and Adam Smith [31]. In previous work on Gregory, I have taken the view that Gregory's concept of sympathy is Humean [22], [23], [24]. The purpose of this paper is to fortify that interpretation in two ways.

First, I shall show that Smith's version of the concept of sympathy is incompatible with Gregory's understanding and use of it in several key respects; whereas Hume's interpretation fits Gregory's text exactly. Second, despite Gregory's having become embroiled in the public dispute between his friend and protégé, James Beattie, and Hume, and despite the controversy surrounding the irreligiosity of Hume, there is strong historical evidence that Gregory was able to distinguish Hume the moral philosopher from Hume the critic of religion. While removing himself at a distance from the latter, he was, I shall document, an admirer of and in philosophical debt to the former. Before turning to these two tasks, I first provide a brief "life" of Gregory and a précis of his medical ethics.

R. Baker, Dorothy Porter and Roy Porter (eds), The Codification of Medical Morality, 145–160.

I. JOHN GREGORY

John Gregory's *Lectures on the Duties and Qualifications of a Physician* first appeared in 1772 [11] in Edinburgh, where he alternated with Collen, as Professor of the Practice of Physic. Born in 1724, Gregory was raised in Aberdeen, where he studied Latin, Greek, Ethics, Mathematics, and Natural Philosophy at King's College, and Edinburgh where he initially undertook medical studies. In 1745, he went to Leyden to continue his studies in medicine. While there, he received, unsolicited, the M.D. degree from King's College in Aberdeen. Having then been elected Professor of Philosophy at King's, Gregory lectured there from 1747–1749, at which time he resigned the position, undertook a grand tour of Europe, and opened a medical practice. In 1754, he removed himself to London to escape the shadow and competition of his older brother, James, a physician in Aberdeen.

In London, Gregory made important friends, including the celebrated Mrs. Montagu. James died in 1755 and Gregory returned to Aberdeen in 1756, where he with his cousin, Thomas Reid and others, including (importantly, for present purposes) James Beattie, founded the Philosophical Society of Aberdeen, also known as the Wise Club. At the meetings of the Wise Club Gregory read a manuscript that later appeared as *A Comparative View of the State and Faculties of Man with those of the Animal World* [9]. In Edinburgh, Gregory moved in distinguished circles, and was an intimate of many of his distinguished contemporaries, including David Hume. In 1773 Gregory died [13]; [30]..

Smellie provides a vivid personal portrait of Gregory:

> In his person, the Doctor was above the ordinary size. His limbs were not very active; he stooped a little in his gait; and his countenance, from a fullness of feature and a heaviness of eye, exhibited no marks of superior powers of mind. But, in conversation, his features became animated, and his eye expressive. He had a warmth of tone and gesture, which were highly interesting. United with this animated manner, he discovered a mildness and simplicity, which were more attractive than the most artificial address. In the company of literary men, his conversation flowed with ease; and, on whatever subject, he delivered his sentiments without affectation or reserve. His benevolent affections were strong, and, in the line of his profession, they were most peculiarly exerted ([30], pp. 116–117).

Smellie adds that Gregory was generous with his students, to whom he was "a friend of easy access, and always ready to assist them with advice and protection" – a standard for medical faculty that might well be worth using as a standard against which to measure ourselves today. Smellie's description concludes with these words:

Upon the whole, whether he is considered a man of genius and of the world, or with regard to his conduct in the line of his profession, few human characters will be found to equal that of the late Dr. JOHN GREGORY ([30], p. 118 emphasis original).

II. A PRÉCIS OF GREGORY'S LECTURES

Gregory's *Lectures* were prompted by the appearance two years earlier of a student version of lectures that he had been giving for several years at the beginning of his regular series of lectures in the Practice of Physic. This version appeared in 1770 as *Observations on the Duties and Offices of a Physician ...* [10]. In the "advertisement" to the "new edition" of his lectures, Gregory writes:

The following Lectures have been read in the University of Edinburgh for several years past; and, as many transcripts of them were, from time to time, taken by my pupils, one of them found its way to the press in the negligent dress in which they were first exhibited. The public, however, having been pleased to afford them a favourable reception even in that form, I thought it a piece of justice I owed to their candour, to give them a thorough revisal, and to make them, as far as I was able, more worthy of its acceptance. This I have now done ([11], p. iii).

Gregory begins his *Lectures* with a definition of "the *practice of medicine*, by which I understand, the art of preserving health, of prolonging life, and of curing diseases" ([11], p. 2, emphasis original). He goes on to address "what kind of genius, understanding and temper naturally fit a man for being a physician, ... the moral qualities to be expected from him in the exercise of his profession ... the decorums and attentions peculiarly incumbent on him as a physician, ... [and] ... that course of education which is necessary for qualifying a physician to practice with success and reputation ..." [11].

In the course of the *Lectures* Gregory addresses the following ethical obligations: to maintain confidentiality ([11], pp. 26–27); to take seriously the suggestions of patients about their own care ([11], pp. 33–34), [reflecting Gregory's interest in "laying medicine open," against, apparently, Cullen's views to the contrary [21]; to be truthful to patients when their "real situation" is "dangerous" ([11], pp. 34–35); not to abandon dying patients and hopeless cases ([11], pp. 35–36); to put patients first during consultation ([11], pp. 38–41); and to dress and behave in a way that is proper and befitting a physician ([11], pp. 51–53), no small issue of mere etiquette [24], especially at the end of the age of manners [6].

Gregory's *Lectures* appeared in several editions in Britain [18], notably one edited by his son, James [16], and the United States [17]. The *Lectures* were also translated, appearing within two decades in German [12] and Italian [14], [15]. Gregory was an acknowledged influence on later medical ethics, including the work of Gisborne [8], Percival [28], and Rush [29], as well as, through them, on the fledging efforts of the American Medical Association (1846–1847) in codifying medical ethics [1].

III. THE SOURCE OF GREGORY'S CONCEPT OF SYMPATHY

All of the duties or obligations that Gregory identifies he founds on Hume's concept of sympathy. The following passage appears early in his *Lectures*:

> I come now to mention the moral qualities peculiarly required in the character of a physician. The chief of these is humanity; that sensibility of heart which makes us feel for the distresses of our fellow-creatures, and which of consequence incites in us the most powerful manner to relieve them. Sympathy produces an anxious attention to a thousand little circumstances that may tend to relieve a patient; an attention which money can never purchase: hence the inexpressible comfort of having a friend for a physician. Sympathy naturally engages the affection and confidence of a patient, which, in many cases, is of the utmost consequence to his recovery. If the physician possesses gentleness of manners, and a compassionate heart, and what Shakespeare so emphatically calls "the milk of human kindness," the patient feels his approach like that of a guardian angel administering to his relief: while every visit of a physician who is unfeeling, and rough in his manners, makes his heart sink within him, as at the presence of one, who comes to pronounce his doom. Men of the most compassionate tempers, by being daily conversant with scenes of distress, acquire in process of time that composure and firmness of mind so necessary in the practice of physick. They can feel whatever is amiable in pity, without suffering it to enervate or unman them. Such physicians as are callous to sentiments of humanity, treat this sympathy with ridicule, and represent it either as hypocrisy, or as the indication of a feeble mind. That sympathy is often affected, I am afraid is true; but this affectation may be easily seen through. Real sympathy is never ostentatious; on the contrary, it rather strives to conceal itself .... The insinuation that a compassionate and feeling heart is commonly accompanied by a weak understanding and a feeble mind, is malignant and false. Experience demonstrates, that a gentle and humane temper, so far from being

inconsistent with vigour of mind, is its usual attendant; and that rough and blustering manners generally accompany a weak understanding and a mean soul, and are indeed frequently affected by men void of magnaminity and personal courage, in order to conceal their natural defects ([11], pp. 19–21).

Gregory first invokes 'humanity' and then 'sympathy'. Both terms are used by Hume and by Smith. For Hume sympathy is a fundamental feature of our moral psychology, our human nature, and activates in us the very same feeling that another person has. We first form an idea of that person's character which leads eventually to an impression in us that matches exactly the feeling that the other person experiences. We are then prompted to act on that feeling. Thus, when we by sympathy experience the feeling of distress that the patient experiences, we are moved to act to relieve that distress, a species of pain, in favor of its relief, which is a species of pleasure.

Smith's moral sense theory varies slightly – but crucially for present purposes – from that of Hume. In a recent commentary on the concept of sympathy, particularly in Hume's philosophy, P. Mercer draws the distinction between Hume's and Smith's understanding of sympathy:

Although, like Hume, he [Adam Smith] thinks that to sympathize with another is to come to have the same feelings as this other person has, Smith does not conceive the process by which this happens in Hume's mechanical terms. Whereas Hume held that sympathy consists in the idea of an emotion being converted into the emotion itself through the enliving association with the impression of self, according to Smith sympathy involves imagining oneself in the other person's situation and thus, in one's imagination, going through all of the emotional experiences he would be going through. We change places 'in fancy with the sufferer' ([25], p. 85).

The difference between Hume and Smith can be put in the following terms. For Hume, for one individual, A, to sympathize with another individual, B, is for A to *experience* the same emotion, E, that B experiences. Hume employs a medical example:

Were I present at any of the more terrible operations of surgery, 'tis certain, that even before it begun, the preparation of the instruments, the laying of the bandages in order, the heating of the irons, with all the signs of anxiety and concern in the patients and assistants, wou'd have a great effect on my mind, and excite the strongest sentiments of pity and terror ([19], p. 576).

By contrast, for Smith, when A sympathizes with B, A *imagines* E. For Hume, A and B share the same emotion, whereas for Smith there need be no emotion E common to A and B, since A is at a remove from E by an act of imagination. For Smith, sympathizing is more cognitive, while for

Hume it is visceral, emotive. So, too, for Gregory, for whom there is a "sensibility of heart which makes us feel for the distresses of our fellow creatures and which, of consequence, incites in us the most powerful manner to relive them." This is direct sympathy, à la Hume, not the indirect – by way of imagination – sympathy of Smith.

Going beyond Hume, Gregory holds a feminist concept of sympathy. Indeed, near the end of the passage above from his *Lectures* Gregory defends sympathy as a virtue appropriate to the "genius" of the physician, while noting that sympathy will not "unman" the physician or make him of "feeble mind". By contrast, Smith distinguishes humanity from generosity: "Humanity is the virtue of a woman, generosity of a man" ([31], p. 190). Smith's sexist views are, indeed, antithetical to those expressed elsewhere by Gregory:

> You will see, in a little treatise of mine just published, in what an honourable point of view I have considered your sex; not as domestic drudges, or the slaves of our pleasures, but as our companions and equals; as designed to soften our hearts, and polish our manners; and, as Thomson finely says:
>
> To raise the virtues, animate the bliss,
> And sweeten all the toils of human life ([13], p. 103).

On this score Gregory and Smith are far apart. Smith's account of humanity contrasts in yet another way with Gregory's use of the term. In his *The Theory of Moral Sentiments* Smith characterizes humanity in the following terms:

> Humanity consists mainly in the exquisite fellow-feeling which the spectator entertains with the sentiments of the persons principally concerned, so as to grieve for their sufferings, to resent their injuries, and to rejoice at their good fortune ([31], pp. 190–191).

For Smith humanity is one of the virtues of indirect sympathy. According to Hume in the *Enquiries*, the sentiment of morals and that of humanity are the same.

> The same endowments of mind, in every circumstance, are agreeable to the sentiment of morals and to that of humanity; the same temper is susceptible of high degrees of the one sentiment and of the other; and the same alteration in the objects, by their nearer approach or by connextions, enlivens the one and the other. By all the rules of philosophy, therefore, we must conclude, that these sentiments are originally the same; since, in each particular, even the most minute, they are governed by the same laws, and are moved by the same objects ([20], pp. 235–236).

And, of course, the sentiment that "enlivens" ideas is sympathy. Hence, sympathy simply replaces humanity. They are interchangeable terms, unlike Smith's treatment of them. In the above passage, Gregory treats them interchangeably, just as Hume does.

On any account that physicians will "feel for the distresses" of their patients, there is a danger that a physician will be swamped by the emotions he or she experiences. Gregory is well aware of this problem when he writes:

On the other hand, a physician of too much sensibility may be rendered incapable of doing his duty from anxiety and excess of sympathy, which cloud his understanding, depress his spirit, and prevent him from acting with that steadiness and vigor, upon which perhaps the life of his patient in a great measure depends ([11], p. 9).

In calling for "steadiness" Gregory more closely follows Hume than he does Smith. Hume's response to the challenge of "violent passions" is "strength of mind ... the prevalence of the calm passions above the violent ...." ([19], p. 418) Gregory's steadiness directly engages and regulates the violent passions that can sometimes be generated by sympathy. Steadiness in thus a "calm passion." Smith's response is characteristically indirect in an appeal to the "spectator's" point of view:

In such paroxysms of distress, if I may be allowed to call them so, the wisest and firmest man, in order to preserve his equanimity, is obliged, I imagine, to make a considerable, even a painful exertion. His own natural feeling of his own distress, his own natural view of his own situation, press hard upon him, and he cannot, without a very great effort fix his attention upon that of the impartial spectator ([31], p. 148).

The issue of response to violent passions for Smith turns on self-command. Moreover, Smith distinguishes self-command from humanity, whereas Gregory treats steadiness as part of sympathy or humanity. Smith writes:

The situations in which the gentle virtue of humanity can be most happily cultivated, are by no means the same with those which are best fitted for forming the austere virtue of self-command .... In the mild sunshine of undisturbed tranquility, in the calm retirement of undissipated and philosophical leisure, the soft virtue of humanity flourishes the most, and is capable of the highest improvement. But, in such situations, the greatest and noblest exertions of self-command have little exercise ([31], p. 153).

Given the importance that Gregory assigns to humanity = sympathy, the chief virtue of the physician, and to its correlative virtue, steadiness, and given that these virtues are exercised together not in "calm retirement" but only in the clinical thick of things, Gregory's humanity and steadiness simply could not be Smith's humanity and "austere ... self-command."

Interestingly, Gregory's understanding and use of a concept of sympathy is philosophically respectable, measured by a twentieth-century yardstick, not just one from the eighteenth century. Mercer's recent philosophical analysis of sympathy is pertinent here.

> I want to maintain that if it is correct to make the statement '*A* sympathizes with *B*' then the following conditions must be fulfilled:
> (a) *A* is aware of the existence of *B* as a sentiment subject;
> (b) *A* knows or believes that he knows *B*'s state of mind;
> (c) there is fellow-feeling between *A* and *B* so that through his imagination *A* is able to realize *B*'s state of mind; and
> (d) *A* is altruistically concerned for *B*'s welfare ([25], p. 19, emphasis original)

Mercer treats each of these four conditions as a necessary but not sufficient condition for the correct use of the statement '*A* sympathizes with *B*'. Gregory would agree and urge that together the four conditions jointly constitute the sufficient condition for the correctness of the statement, 'Physician *A* sympathizes with patient *B*'.

IV. THE PROBLEM OF HUME'S IRRELIGIOSITY

If the preceding textual analysis is correct, then the best reading of Gregory on sympathy requires a Humean debt and reflects a Humean influence, because the views of the only other (then) contemporary proponent of sympathy cannot fit Gregory's text. There is, however, an historical puzzle that must be solved, for it stands in the way of accepting this interpretation of Gregory's medical ethics.

The historical puzzle begins with James Beattie's attack on Hume, in 1770, for the latter's *irreligiosity* [2]. Beattie (1735–1803) was in his time an important figure, but his stature has since collapsed in the history of philosophy. Hume's very negative reaction to Beattie's attack lead to a crossfire in which Gregory was caught, as Beattie's mentor and friend. It seems, at first, that Gregory sided with Beattie against Hume. If Gregory did indeed do so, how could he be Humean in his medical ethics? This question expresses the heart of the historical puzzle.

A useful starting point for solving this important puzzle is a letter that Gregory wrote to Mrs. Montagu, the prominent social and intellectual figure whom he met while residing earlier in London, after Beattie published his attack on Hume. Beattie's attack came at a time when Hume was at the height of his fame; it caused quite a stir. E. C. Mossner [27], in his biography of Hume, notes that Beattie's book, which appeared in 1770, "was chiefly responsible for disturbing the philospher's tranquility" ([27], p. 577). Indeed, contrary to his usual practice of ignoring his critics, Hume did reply publicly to Beattie. Beattie's aim is portrayed by Mossner in a manner that is clearly sympathetic to the subject of his biography: " ... it was Beattie's intention to arouse the emotional prejudices of his readers" ([27], p. 577). In this context Mossner quotes Gregory's letter from 3 June 1770 to Mrs. Montagu:

> Zeal for his Cause has made him [Beattie] treat Mr Hume sometimes with a degree of Severity which I think had better been spared. I detest Mr Hume's Philosophy as destructive of every principle interesting to Mankind & I think the general spirit that breathes in his History unfavourable both to Religion and Liberty, tho in other respects one of the most animated, entertaining & instructive Historys I have ever read. But I love Mr Hume personally as a Worthy agreeable Man in private Life, & as I believe he does not know and cannot feel the mischief his writings have done, it hurts me extremely to see him harshly used ([27], p. 580).

On the face of it this letter indicates that, while Gregory may have liked, even admired and felt for Hume the person, he rejects altogether Hume the philosopher. As J. Dunn puts it, Gregory "was certainly strongly opposed to the philosophy of Hume ...." ([5], 128). Agnes Grainger Stewart is more pointed in her remarks on this subject. She notes that Beattie and T. Reid

> ... were engaged in combating the teaching of David Hume, which had become very fashionable, and Gregory, though much attached to David Hume as a man, feared him as a teacher, and dreaded the growth of scepticism which marked the time ([32], p. 119).

She adds that "Gregory's mind was deeply religious, but it was of that sort that lives more by meditation than church-going" ([32], p. 121).

A more sustained look at the historical relationship between Gregory and Hume preceding the "Beattie incident," (if I may call it that), calls this prima facie reading into question. That more sustained look begins with evidence that we have about Gregory's earliest acquaintance with the work of Hume, namely, the response of the Wise Club to Hume's *Treatise*. J. H. Burton's *Life and Correspondence of David Hume* [3] includes two letters that are pertinent. The first is Hume to Blair in 1763:

> I beg my compliments to my friendly adversaries, Dr. Campbell, and Dr. Gerard, and

also to Dr. Gregory, whom I suspect to be of the same disposition, though he has not openly declared himself such" ([3], p. 154).

Hume's remarks were passed on to Thomas Reid, a co-founder of the Wise Club, who responded in a letter from "King's College, 18th March, 1763":

Your friendly adversaries, Drs. Campbell and Gerard, as well as Dr. Gregory, return their compliments to you respectfully. A little philosophical society here, of which the three are members, is much indebted to you for its entertainment. Your company would, although we are good Christians, be more acceptable than that of St. Athanasius; and since we cannot have you upon the bench, you are brought oftener than any other man to the bar, accused and defended with great zeal, but without bitterness. If you write no more on morals, politics, or metaphysics, I am afraid we shall be at a loss for subjects ([3], p. 155).

Several features of this letter are worth comment. Reid conveys that Gregory is a friendly adversary to Hume's work, not an unconditional or uniform opponent – although this is conveyed privately, in a letter by another, Reid. In addition, Hume's ethics, political philosophy, and metaphysics are praised. At the same time – importantly, I think – Hume's skepticism regarding religion is not mentioned. This omission is curious, given the provocative nature of what Hume writes, for example, in the *Treatise* on the subject," ... errors in religion are dangerous; those in philosophy only ridiculous" ([19], p. 272).

The comments about Athanasius are especially interesting in this respect. Athanasius, church father and patriarch of Alexandria, condemned the Arian heresy, which was anti-trinitarian in its denial that Jesus was one in substance with God the Father. Reid's comment can be read as the expression of the preference of the Wise Club members for a skeptic who argues over an unquestioning, unreflective believer. In short, Reid's comment seems to constitute a subtle compliment to Hume's skepticism, perhaps even for some of its implications regarding religious beliefs. This embrace of Hume's skeptical philosophy, but not necessarily (all of) its implications for religion, becomes the central theme of the Gregory-Hume connection.

The understated fashion in which the Wise Club managed its simultaneous embrace of Hume and reservations about its philosophical implications for religion was noted by Beattie in a letter of 1770 to a Dr. Blacklock. He refers to Reid and Campbell:

I know likewise that they are sincere, not only in the detestation they express for Mr. Hume's irreligious tenets, but also in the compliments they have paid to his talents;

for they both look upon him as an extraordinary person; a point on which I cannot disagree with them ([7], vol. i, p. 123).

Beattie goes on to wish that Reid and Campbell has gone further in "their researches" and "expressed themselves with a little more firmness and spirit" ([7], vol i, p. 123). Beattie more than made up for their shortcoming in the latter respect.

When Gregory returned to Edinburgh he became a member of Hume's "circle." As Smellie puts it, "In the later period of his life, when he lived in Edinburgh, he lived in habits of great intimacy with most of the Scottish Literati; such as ... David Hume ...." ([30], pp. 117–118). Beattie provides important evidence that Gregory's main objection to Hume concerned his "irreligious tenets"; in a letter Beattie wrote to Mrs. Montagu in which he describes an exchange between Gregory and Hume when the latter was near death:

Yet Mr. Hume must have known, that, in the opinion of a great majority of his readers, his reasonings, in regard to God and Providence, were most pernicious, as well as most absurd. Nay, he himself seemed to think them dangerous. This appears from the following fact, which I had from Dr. Gregory. Mr. Hume was boasting to the Doctor, that, among his disciples in Edinburgh, he had the honour to reckon many of the fair sex. 'Now, tell me,' said the Doctor, 'whether, if you had a wife or a daughter, you would *wish* them to be your disciples? Think well before you answer me; for I assure you, that, whatever your answer is, I will not conceal it.' Mr. Hume, with a smile, and some hesitation, made this reply: 'No, I believe scepticism may be too sturdy a virtue for a woman.' Miss Gregory will certainly remember, that she had heard her father tell this story ([7], vol. ii, p. 35, emphasis original).

Again, Gregory's concern is with the implications of Hume's skepticism for religious beliefs, not with his basic philosophy.

When the Beattie-Hume controversy broke out, Gregory, who was then in Edinburgh, and who was at once Beattie's life-long friend and one of Hume's intimates, found himself in the middle of a ticklish situation, as is evident from a subsequent letter from Gregory to Beattie, dated 20 June 1770. Interestingly, the letter begins with a rebuke of Beattie: "Much woe has your essay wrought me." ([7], vol. i, p. 164) Gregory then goes on to report that Hume was quite angry:

As it was known that the manuscript had been in my hands, I was taken to task for letting it go to press as it stands ([7], vol. i.,p. 164).

Gregory continues with a criticism mainly of Beattie's style, noting that Beattie writes "with a spirit and elegance very uncommon on such subjects ..." and characterizes that style as one of "warmth" ([7], vol. i,

p. 164). Gregory continues in this vein at some length:

I wished, at the same time, some particular expressions had been softened; but denied there being any personal abuse. In one place, you say, '*What does the man mean*?' This, you know, is very contemptuous. In short, the spirit and warmth with which it is written has got it more friends and more enemies than if it had been written with that polite and humble deference to Mr. Hume's extraordinary abilities, which his friends think so justly his due. For my own part, I am so warm, not to say angry, about this subject, that I cannot entirely trust my own judgment; but I really think that the tone of superiority assumed by the present race of infidels, and the contemptuous sneer with which they regard every friend of religion, contrasted with the timid behavior of such as should support its cause, acting only in the defensive, seems to me to have unfavourable influence. It seems to imply a consciousness of truth on the one side, and a secret conviction, or at least diffidence, of the cause, on the other. What a difference from the days of Addison, Arbuthnot, Swift, Pope & Co. who treated infidelity with a scorn and indignation we are strangers to. I am now persuaded the book will answer beyond your expectations. I have recommended it strongly to my friends in England ([7], vol. i, pp. 164–165, emphasis original).

Notice again, it is only the implications for religion of Hume's skepticism that fall under Gregory's criticism. Moreover, he puts his remarks in an interesting context, focusing on "infidels," among whom he does not list Hume. Moreover, he immediately moves to a critique of those who have attacked Hume – among whom Gregory was not included. The last two sentences seem to be ironic, at least.

The controversy over Beattie continued to simmer, and Gregory once again writes to Beattie on the subject on 26 November 1771. Gregory suggests to Beattie that he opposes Hume, not because the latter is a "bad metaphysician, but because he has expressly applied his metaphysics to the above unworthy cause" ([7], vol. i, p. 180). Earlier in the letter this "unworthy cause" is characterized in the following terms:

You write with warmth against him [Hume], because he has endeavored to invalidate every argument brought to prove the existence of a Supreme Being; because he has endeavored to invalidate every argument in favour of a future state of existence; and because he has endeavored to destroy the distinction between good and evil ([7], vol. i, p. 180).

Later in the letter Gregory adds:

But I have never heard that he, or any of his friends, have pretended, that you do him injustice in these respects ([7], vol. i, p. 181).

These passages make it even clearer that Gregory is not just distinguishing the "man," Hume, from his philosophy, but Hume's metaphysics, ethics, and, skeptical method from one area of its application. He also chastises Beattie for an intemperate approach to philosophical criticism. Commenting on the above letter, Forbes, Beattie's biographer, writes:

... Dr. Gregory has placed in the most proper point of view the accusation brought by the friends of Mr. Hume against Dr. Beattie, of having, in his Essay on Truth, treated the principles of the sceptical philosophy with too much asperity ([7], vol. i, p. 179).

Interestingly, in his earliest biography of Hume, Mossner makes a similar point – which disappears, as near as I can tell, in his masterful, later biography of Hume.

The spirit of tolerance illumined the Edinburgh circle. Though many of the literati were clergymen, all were enlightened. To them, the Great Infidel presented no insurmountable problem of ethics. While they reprobated Hume's religious opinions, they were yet able to distinguish between the opinions and the man; if they abhorred the first, they loved the second ([26], p. 198).

It may well be that, in the enlightened view of members of his "circle," Hume's main threat was not to religion per se, but rather to *particular religious beliefs* and to particular religions that are based upon those beliefs. Calderwood, a biographer of Hume, provides an enlightened account of Hume's views that his "circle" might have approved:

These extracts [of Hume's texts] shew how clearly Hume maintained his conviction of the evident value of religion, even when tracing the inconsistencies which appear in its history among the several nations and tribes of men. His mental characteristics, intellectual and emotional, induced him to treat scornfully of these inconsistencies, as if they were traces of hyprocrisy. This tendency appeared so offensively in the first volume of his *History of England* as to subject him to sever criticism ([4], p. 102).

In summary, the evidence found in the sources considered here supports the conclusion that Gregory not only embraced Hume the man, he also clearly rejected only Hume's *religious* skepticism, and mainly that part of it taken up by ill-mannered followers of Hume. Gregory did not reject the central elements of Hume's philosophy, including those of his ethics. Indeed, he seems to have respected them and seems also to have thought that Beattie's attack did not reach them and that, moreover, it was not intended to do so. The historical puzzle seems, therefore, solved: on the basis of the available historical evidence, there are no grounds to exclude a Gregory-Hume connection on the subject of sympathy, the central concept of Hume's version of common sense-based ethical theory.

V. CONCLUSION

I have endeavored to provide further evidence for reading Gregory's medical ethics in Humean terms. David Hume invented the philosophical concept of sympathy and Adam Smith developed the major alternative variant of the concept. Gregory knew both men and their philosphical work. Textual exigesis provided here supports the conclusion that only Hume could be the source of the concept of sympathy in Gregory's *Lectures* on medical ethics. Careful consideration of relevant historical materials shows that there is no puzzle involved in this interpretation. Perhaps relying on his earlier philosophical training, Gregory was able to distinguish Hume the moral philosopher from Hume the religious skeptic. In any event, Gregory relied on and exploited the best philosophical work of his time in formulating his medical ethics. He is, in my judgment, the first modern medical ethicist to be identified in the Anglo-American literature.

ACKNOWLEDGEMENTS

I want to thank the Center for Ethics, Medicine, and Public Issues of the Baylor College of Medicine, Houston, Texas, USA, and the Wellcome Institute for the History of Medicine, London, England, for their support to allow me to attend the conference at the Wellcome Institute in December, 1989, at which an earlier version of this paper was presented. I am indebted to the editors of this volume for their suggestions, especially to Robert Baker for pushing me, correctly, into an unequivocal commitment to the view that Gregory's concept of sympathy – and thus his medical ethics – are thoroughly Humean in character.

Baylor College of Medicine
Houston, Texas, USA

BIBLIOGRAPHY

1. American Medical Association: 1846–1847, 'Code of Medical Ethics', *Proceedings of the National Medical Convertion*, American Medical Association, Chicago, pp. 83–106.

2. Beattie, J.: 1976, *Elements of a Moral Science*, A facsimile reproduction [of the 1770 edition] with an introduction by J.R. Irvine, Scholars' *Facsimiles and Reprints*, Delmar, New York.
3. Burton, J.H.: 1846, *Life and Correspondence of David Hume*, William Tail, Edinburgh.
4. Calderwood, H.: 1898, *David Hume*, Oliphant Anderson and Ferrier, Edinburgh and London.
5. Dunn, J.: 1964, 'Authorship of Gregory's Critique of Hume,' *Journal of the History of Ideas* 25, 128–129.
6. Fissell, M.: 1991, 'Medical Ethics or Polite Behavior: Some Controversies from the South-West', in this volume, pp. 15–45.
7. Forbes, Sir W.: 1824, *An Account of the Life and Writing of James Beattie, LL. D. Late Professor of Moral Philosophy and Logic in the Marischal College and University of Aberdeen. Including many of his Original Letters*, E. Roper, London.
8. Gisborne, T.: 1794, *An Enquiry into the Duties of Men in the Higher and Middle Classes of Society in Great Britian, Resulting from their Respective Stations, Professions, and Employments*, B. and J. White, London.
9. Gregory, J.: 1765, *A Comparative View of the State and Faculties of Man with those of the Animal World*, J. Dodsley, London.
10. Gregory, J.: 1770, *Observations on the Duties and Offices of a Physician, and on the Method of Prosecuting Enquiries in Philosophy*, W. Strahan and T. Cadell, London.
11. Gregory, J.: 1772, *Lectures on the Duties and Qualifications of a Physician*, W. Strahan and T. Cadell, London.
12. Gregory, J.: 1778, anonymous (trans.), ... *Vorlesungen über die Pflichten und Eigenschaften eines Artes. Aus dem Englischen nach der Neuen und Verbesserten Ausgabe Übersetzt*, Caspar Fritsch, Leipzig.
13. Gregory, J.: 1786, *A Father's Legacy to his Daughters to which is prefixed an account of the life of the Author*, A. Strahan and T. Cadell, London, W. Creech, Edinburgh.
14. Gregory, J.: 1789, F. F. Padovano (trans.), *Lezioni Sopra i Doveri e le Qualita di un medico* ..., Gaetano Cambiagi, Firenze, Italy.
15. Gregory, J.: 1795, F. F. Padovano (trans.), *Lezioni Sopra i Doveri e le Qualita di un medico*, Baldassare Comino, Pavia, Italy.
16. Gregory, J.: 1805, *Lectures on the Duties and Qualifications of a Physician. Revised and corrected by James Gregory, M.D.*, W. Creech, Edinburgh, and T. Cadell and W. Davies, London.
17. Gregory, J.: 1817, *Lectures on the Duties and Qualifications of a Physician*, M. Carey and Son, Philadelphia, Pennsylvania.
18. Gregory, J.: 1820, *On the Duties and Qualifications of a Physician. New edition*, J. Anderson, London.
19. Hume, D.: 1968, L. A. Selby-Bigge (ed.), *A Treatise of Human Nature*, The Clarendon Press, Oxford, England.
20. Hume, D.: 1988, P. H. Nidditch (ed.), *Enquiries Concerning Human*

Understanding and Concerning the Principles of Morals, The Clarendon Press, Oxford, England.
21. Lawrence, C. J.: 1975, 'William Buchan: Medicine Laid Open', *Medical History* 19, 20–35.
22. McCullough, L. B.: 1978, 'Historical Perspectives on the Ethical Dimensions of the Patient-Physician Relationship: The Medical Ethics of Dr. John Gregory', *Ethics in Science and Medicine* 5, 47–53.
23. McCullough, L. B.: 1983, 'The Legacy of Modern Anglo-American Medical Ethics: Correcting Some Misperceptions', in E. Shelp (ed.), *The Clinical Encounter*, D. Reidel Publ. Co., Dordrecht, the Netherlands, pp. 47–63.
24. McCullough, L. B.: 1984, 'Virtues, Etiquette, and Anglo-American Medical Ethics in the Eighteenth and Nineteenth Centuries', in E. Shelp (ed.), *Virtue and Medicine*, D. Reidel Publ. Co., Dordrecht, the Netherlands, pp. 81–92.
25. Mercer, P.: 1972, *Sympathy and Ethics: A Study of the Relationship between Sympathy and Morality with Special Reference to Hume's Treatise*, The Clarendon Press, Oxford, England.
26. Mossner, E. C.: 1943, *The Forgotten Hume: Le bon David*, Columbia University Press, New York.
27. Mossner, E. C.: 1980, *The Life of David Hume*, 2nd ed., The Clarendon Press, Oxford, England.
28. Percival, T.: 1803, *Medical Ethics, or a Code of Institutes and Percepts Adapted to the Professional Conduct of Physicians and Surgeons*, S. Russell, Manchester, England.
29. Rush, B.: 1789, *Observation on the Duties of a Physician and the Methods of Improving Medicine, Accomodated to the Present State of Society and Manners in the United States*, Prichard and Hall, Philadelphia, Pennsylvania.
30. Smellie, W.: 1800, *Literary and Characteristical Lives of John Gregory, M.D., Henry Home, Lord Kames, David Hume, esq., and Adam Smith*, Alex Smellie, Edinburgh.
31. Smith, A.: 1976, D. D. Raphael, A. L. MacFie (eds.), *The Theory of Moral Sentiments*, The Clarendon Press, Oxford, England.
32. Stewart, A. G.: 1901, *The Academic Gregories*, Oliphant Anderson and Ferrier, Edinburgh and London.

CHAPTER 7

JOHN V. PICKSTONE

THOMAS PERCIVAL AND THE PRODUCTION OF MEDICAL ETHICS

My interest in Thomas Percival arises from my position as a historian of medicine in, and indeed from, the Manchester region. Percival of course is a local hero. He is also an important figure in world medicine, and so Manchester medicine in the late eighteenth century has been the subject of major studies by historians in America [52], [14], [56]. And we have two newer reasons for re-examining Percival and his context.

One is the recent work on the politics of Manchester medicine.* The second reason, newish in Britain, if not so new in the States, is the contemporary interest in problems in medical ethics.

My aim then is to situate Percival's medical ethics – in his life and career, in Manchester medicine, in the history of medical occupations in England, and in the social history of the industrial revolution. I hope I can add new perspectives, not just on the intimate issues of Percival's development and beliefs, nor just on the extraordinary context of his work in the 1790s, but on how these all fitted into the wider development of medicine and industrial society.

First, I shall briefly survey some of the secondary literature on medical ethics, to find the general approaches and suggestions there as to why Percival published his *Medical Ethics* in 1803. Then I shall switch to Percival's biography and focus especially on his work as a public moralist. Third I shall review the history of medicine in Manchester between 1784 and 1803, to see where the *Ethics* came from. And, finally, I shall explore the general question – what was Percival's project in this work, and how are we to understand its production at this particular place in historical time?

There seem to be two main genres of secondary literature about medical ethics. In the first, exemplified, for example, by Waddington on early nineteenth-century Britain, and by Berlant on America, medical ethics reflected in professional claims to status, or not much more [59], [60], [4]. In the second genre, well exemplified by Ludwig Edelstein on Greek medicine, medical ethics reflect major, general ethical positions in

R. Baker, Dorothy Porter and Roy Porter (eds), The Codification of Medical Morality, 161–178.

the society of which doctors formed but a small part [11]. Both approaches have their virtues, and a history of medicine which is adequately social can surely encompass both.

From the French Revolutionary Wars through the mid-nineteenth century, medical reform was controversial and professional relations within medicine were confused and contested. Particularly problematic was the division between consultants and general practitioners – consultants had (charitable) hospital practices, they gave second opinions on the cases of other doctors, but they often also acted as practitioners of first resort. Under such conditions, better internal policing of medical occupations was required to reduce overt conflict and so improve the public image of doctors. A better image, a claim to unity, would facilitate the granting of state protection, etc. Hence the uses of ethical codes [60].

But we should be wary of equating too closely the nineteenth-century uses of Percival with his own intention and situation in the 1790s when he first composed the work. Indeed, a nice measure of that distance is offered by the book which is usually referred to as a second British edition of Percival's medical ethics, published anonymously in 1827 [42]. It is by a west-country practitioner who used a severe abridgement of Percival's text as a vehicle for his own extended comments on medical education, reforms, and professional hierarchies (and the ways in which evangelical laymen chose co-religionists as practitioners, thus undermining doctors of true professional merit). That kind of reform tract (if not the point about evangelicals) was common by the 1820s; Percival's own text – a set of professional rules – was something rather different, and rarer (at least in Britain).

We may agree with Waddington that the problematic structures of medicine around 1830 called forth many pamphlets on professional organization and reform, but it is not at all clear that works on medical *etiquette* were a major feature of this reformist literature, still less that Percival's text was much used. Waddington suggests that there were quite a lot of works on medical ethics in early nineteenth-century Britain, but he does not give many examples. Percival's first edition is a rare book; and with the exception of that included in the *Life and Works*, there was no proper new edition until 1849 [43].

The case may be different for the United States. There, it seems, Percival's rules were widely used by local medical societies and then by the American Medical Association. Had Percival not existed, one may argue, Americans would have had to invent him. But the case is much less clear for the reformers of British medicine.

What then shall we say about 'need' and 'response' in late eighteenth-century Britain? Historians tend to suggest that the works of Percival or Gregory were responses to professional problems of general impact [4]. But again there is room for doubt. It is not easy to find medical ethics texts other than those of Gregory or Percival. There was undoubtedly much moralizing about medicine, and perhaps new professional arrangements accentuated that common tendency, but only in one or two, perhaps peculiar, circumstances did medical men sit down to produce works specifically on medical ethics. Perhaps, then, we should refine our question about the circumstances which called forth "ethics." We should focus the question and ask, for Gregory and for Percival, about the specific historical circumstances which stimulated and shaped their texts.

As Christopher Lawrence has maintained, Gregory wrote to defend what he took to be the traditional values of Scottish medicine (and its common-sense philosophy) *against* the skepticism of the Scottish enlightenment [27]. Hence the paradox that Gregory has since appeared as a radical. His appeal that broadly educated gentlemen be judges of medicine was, in fact, a conservative defence of a broad God-fearing culture which clever Edinburgh physicians seemed, to him, to scorn. I wish to develop a similar argument for Percival and late eighteenth-century Manchester. Percival, too, represents a rather specific and peculiar conjunction, not just at the personal level, but at the level of Manchester and British history.

Let us begin with the man in his milieu, not in the 1790s but the 1770s, when he was newly secure at the peak of the local profession. Percival had come from Warrington, a small manufacturing town and river-port on the Mersey, about twenty miles to the west of Manchester. He was the son of a merchant, grandson of a surgeon-apothecary, and nephew of a physician. Warrington is known to historians as the site of the Dissenting Academy, a college for students and would-be ministers, maintained by a small group of Unitarians (the rationalist, theistic tendency in eighteenth-century English presbyterianism). Joseph Priestley taught at Warrington Academy; Thomas Malthus was a pupil (though an Anglican). Percival was the Warrington Academy's first student, leaving just after Priestley arrived [18], [35].

From Warrington, Percival went to Edinburgh University, where he became a member of the Medical Society in 1763. By 1765, aged 25, he was already well known in medical and intellectual circles. In that year he travelled on the continent and acquired a Leiden MD; he was also elected a Fellow of the Royal Society of London – hardly the intellectual honor that such an election would become, but a mark of considerable intellectual and social respectability, especially in one so young.

For two years he practised medicine in his home town. In 1767 he decided to move to Manchester. London had been his first choice, but the death of his patron there, Lord Willoughby de Parham, had deprived him of his point of entry. In Manchester he had good family and personal connections, non-aristocratic to be sure, but significant figures in the capital of a prosperous region, increasingly dominated by the spinning and weaving of wool and linen cloths. He quickly established himself. Dr Percival, wags said, would perceive all and receive all [18].

In 1750, Manchester's population was about 25,000; by Percival's death in 1804 it would exceed 100,000. In 1750, the town had two churches and two dissenting chapels, one of them Unitarian. As in other provincial towns the local bourgeoisie were becoming self-consciously polite, making their own public entertainment in assembly rooms and such; subscribing for good causes such as street improvements. An infirmary (a charity hospital) was begun in 1752 by the local gentry and a young local surgeon, Charles White, fresh from his surgical training in London with the Hunter brothers. It developed unusually well, adding a lunatic hospital in the 1760s and later a set of public baths, not so much for the patients as for the use of middle-class subscribers, for whom Turkish or cold baths were a handy, cheaper alternative to local spas such as Buxton [45].

That sort of urban society was commonplace in Britain, but some aspects of Manchester were not. In 1761, for example, the Duke of Bridgewater's canal was opened, to bring coal into Manchester from the Duke's coal mine in nearby Worsley. It exemplifies the involvement of aristocracy in local society (and the industrial revolution), and the culture of commercial and technical inventiveness which was then characteristic of provincial Britain. But the canal can also be a context for Percival, for in one of his works he described a family visit paid to the canal wharves, probably around 1770 [36]. As Percival approached the canal with several of his children, a donkey was pulling a cart up the slope from the wharf. The cart was loaded with coal, the donkey was slipping and the donkey driver was beating it. Euphronius, for that is how Percival presented himself in his moral tales, remonstrated with the donkey driver – to no avail. His son, young Jacobus, however, had the pragmatic wit to offer a few pence to reinforce the lesson.

Down by the wharf, father enjoyed pointing out the various products – perhaps reminded of his own school days at Warrington Academy, where science had often been so taught [49]. But again the family was disturbed; again the culprits were donkey-drivers, now laughing and swearing as they tossed halfpennies – would they spend them on food for their donkeys, or on drink for themselves? Euphronius would have liked to remonstrate, but felt threatened by possible responses. So the family withdrew, to discourse on man's duty to animals, especially to such serviceable animals as donkeys. There was much to be learned from animals in the world of Euphronius; much to be passed on, if reason and passion were to be so joined in the young, that wisdom and sensibility would increase and virtue flourish.

Those who know something of Percival's writings, or even of popular eighteenth-century genres, will have recognized a moral tale. It appeared in Percival's *A Father's Instructions to His Children; consisting of moral tales, fables and reflections, adapted to different periods of life, from youth to maturity, and designed to promote the love of virtue, a taste for knowledge, and an early acquaintance with works of nature* (Part one, London, 1775). Percival wrote three volumes of *Father's Instructions* [37], [38]. That he did so tells us much about him, his position, and his claims to a public. And before we come to Percival's medical pre-occupations it is worth spending a few moments on the exact place where he wrote his first volume – a new country house to the north of the city [18].

His friends, both professional and mercantile, had similar retreats. The Phillips family, Unitarian mill-owners, had a large estate at Prestwich [15]; Thomas Butterworth Bayley, the magistrate and Percival's friend since student days at Edinburgh, lived out towards Swinton [3], [40]. Charles White, Percival's surgical equivalent, had a very large house in town, but also a country retreat at Ashton-upon-Mersey where he collected new species of trees [6]. Theirs was an urban culture inasmuch as they worked in town and socialized there, especially in winter, but this culture had its rural aspects. And it was in the countryside that a gentleman could best find that self which could be passed on to his children [10].

In writing *Father's Instructions*, Percival, in a sense, was projecting this family scene to his counterparts elsewhere. This was not the professional interchange of his medical papers or of *Essays Medical and*

Experimental [35]; the audience was not restricted to doctors or devotees of natural philosophy. The readers of *Father's Instructions* were gentlemen and their families in the higher and middle classes of society. Percival was now rich enough to own a country house; he was sufficiently well-connected to see himself as a representative of English culture; he could be a gentleman as well as private tutor, secular cleric as well as natural philosopher. Cultured fathers everywhere could learn to be educators; to be Aristotles to their own little Alexanders. This was suburban Manchester, to be sure, but it was also deeply neo-classical, not least in the correspondences presumed between the natural, the social, and the moral [33].

From the *Father's Instructions* and similar writing, it is not difficult to judge the ethical tendency to which Percival belonged. He stood in the tradition of Bishop Butler, the then celebrated exponent of *virtue ethics*. This matters. Percival was *not* a utilitarian, though Jeremy Bentham was influential before Percival died; he was not a materialist of the school of David Hartley or Joseph Priestley, for all his Unitarian connections. Nor was he an evangelical. He is best linked with a saintly, tolerant Anglican divine [18], [30]. We note here that Percival was a convert from the Church of England to Unitarianism, and that Percival's eldest son became an Anglican clergyman, without, it would seem, any complaint from the father. We may also note that Butler, bishop of Bristol and then of Durham, had attended a Dissenting Academy.

Butler came to be known as a simple man's Kant, but he was sufficiently serious a moral philosopher to have attracted a recent scholarly monograph [34]. He stood for the defence of Christian religion against skepticism and secular rationalism, not by setting religious ethics *against* secular ethics (as would later become popular), but by accepting the possibility of a secular ethics to which religious belief was superadded. Butler's secular ethics – the supposed product of man's intrinsic moral sense – was largely classical in its derivation, largely *stoical*. Here one sees the resonance of ethical discussion with the classical bent of so much eighteenth-century culture – from training in Latin, to classical architecture, to notions of civic virtue. The key was the concord of the laws of nature, of society and of God. Man was moral when true to his own *nature*. The nature of virtue and the virtues of nature were to be studied from childhood. How then did virtuous physicians behave in private practice and in public politics?

By the 1780s, Percival was the acknowledged leader of local culture, a figure of national significance, later a correspondent of such cosmopolitan figures as Benjamin Franklin. He was founder of the Manchester Literary and Philosophical Society, a sponsor of Manchester New College (the successor to Warrington Academy), and a kind of super consultant to the Manchester Infirmary. He was a leading member of the Congregation of Cross Street Unitarian Chapel, which included much of Manchester's burgeoning mercantile wealth. He was in a position now to be paternalist not just to his children (or via books to those of his peers); he was equipped and inclined to be a city father, a public paternalist, solving public problems rather larger than donkey-driving by the Bridgewater canal.

We come then to the particularities of Manchester medicine and to the circumstances which prompted the instructor of children to become the instructor of his professional fellows. We could begin with 1784 and the typhus epidemic at Robert Peel's spinning factory at Radcliffe, near Bury. (This is Robert Peel the elder, a yeoman turned industrialist who became a Tory MP; his well-educated son became Prime Minister). Fever in Lancashire in the 1780s was political because spinning mills were new and very contentious. Earlier attempts to build mills had provoked riots; by the 1780s, Arkwright's water-frames were being installed under license in huge factories. Workers believed that the mills generated fever, as did ships, army camps, and jails. The magistrates in 1784 asked Percival and other Manchester physicians to report on the Radcliffe outbreaks [9], [32], [46].

Out came the favored remedies and prophylactics of the medical enlightenment: cleanliness and ventilation, white-washing houses and boiling clothes. It was not clear whether fever was in fact generated in mills, but Percival was concerned about the employment of pauper apprentices, especially on night work. Robert Peel was not pleased by the public criticism and controversy, though years later, in different political conditions, he was to use Percival's recommendations as a basis for the 1803 Health and Morals of Apprentices Act, the ineffective precursor of nineteenth-century factory legislation [44].

Generally the 1780s were an ecumenical period for the Manchester political élite. They stood more or less together in agencies such as the new Sunday Schools, which sought to civilize the children of the expanding working class. But by the end of the decade, the growing problem of fever in the city, together with a raft of contentious political issues, had divided the town. The Tories controlled most of the rudimentary statutory agencies – the parish and poor-law relief. Dissenters concentrated on new voluntary agencies, the Literary and Philosophical Society, and the associated attempts at organizing higher education.

The politico-religious divide was exacerbated by the campaign to abolish slavery, which directly affected the region's main port – Liverpool. Percival was anti-slavery, like most of the dissenters, who also campaigned against their exclusion from certain civic privileges. All these issues helped polarize the town and provided a context for disputes about health and hospitals. Responses to the French Revolution added to dissension, which by 1790 was institutionalized: a Church and King club stood opposed to a Constitution Club. (Shortly afterwards there was a Society for the Putting Down of Levellers.) By 1791–92, as war with France loomed, hostility would become violent [15], [24], [46].

By 1790, the Infirmary was already contentious. It was a voluntary agency and its government was open to a large number of subscribers, so it was formally democratic; but its day-to-day business was controlled by the surgeons, or surgeons-turned-physicians, notably the White family and the Halls, who were all Tory Anglicans. The details of this remarkable contest are not required here, but its outlines are important for Manchester medicine in the 1790s. As is generally known, the hospital *revolution* was the immediate spur to the first version of Percival's medical ethics, written in 1793–94, particularly to the chapters on the etiquette of hospital practice.

The dispute was primarily about whether or not the honorary staff complement of the Infirmary should be doubled from six to twelve (i.e., to six physicians and six surgeons). The proposed increase was intended to enlarge the out-patient and home-patient role of the hospital, including the domiciliary care of fever patients. The reformers and expansionists, who were mostly dissenting Whigs and physicians, wished to take over the hospital, extend its public health role, and reduce the hold of the key local surgeons over medical practice in Manchester. They were disturbing a local medical hierarchy which had held for twenty years or more.

Physicians had been relatively few and it was not difficult for them to secure Infirmary posts. (Percival, after a brief hospital appointment, had retired to private practice and been made physician extra-ordinary). Surgeons and surgeon-apothecaries were much more numerous, but they were nicely divided into a large group with few pretensions, and a much smaller élite group, including the Whites and the Halls. Charles White was well known nationally and could charge large apprenticeship fees; his big house in town was much frequented and contained an extensive museum. He seems to have made sure that his ex-apprentices left town so as not to create competition. The situation was well understood, and accepted by some, for indeed, an infirmary open to all local practitioners could seem a very republican experiment. But perhaps it was going too far when White's son Thomas, who had trained with his father as a surgeon and then taken a medical degree, was elected to one of the physician's posts in 1786. Certainly John Ferriar, a young Scottish physician and a protegé of Percival, was anxious to obtain an Infirmary post; he could not have relished the preferment of Thomas White [61].

It was two radical friends of Ferriar – Thomas Walker (later Borough Reeve) and Thomas Cooper – who began the agitation to increase the staff. The reform party lost the first round in 1788–89, but by 1790 the town had become increasingly worried about fever and increasingly polarized over political issues. In a very large meeting of subscribers, the reformers got their way. The Whites and the Halls were left to run a small maternity charity which they had begun as part of this dispute. They did what they could with what remained to them; they tried to teach medicine at the maternity hospital and they had the support of Tory and military interests. For a while then, Manchester medicine had two rival institutional bases [46]. The dispute, moreover had brought several new doctors into town; after two decades of stable professional hierarchy, the situation had become much more fluid and contentious [63].

Percival had been prominent in the Infirmary dispute and he remained so as the Infirmary was renovated and extended. Over the next few years he helped draft a new set of hospital regulations, including his first attempt at medical ethics. He was asked by the physicians and surgeons of the Infirmary to draw up a code of laws to regulate hospital practice. This was to be incorporated into the Infirmary rules. Hence the aphoristic style which proved so useful.

But that was not the limit of Percival's ambitions. By 1794 he had drafted his *Medical Jurisprudence*, which he circulated to a galaxy of medical and literary friends. This version included rules for private as well as public practice. It is worth enquiring why he expanded the project in this way [39].

First, we remind ourselves that Percival was already a widely-read *moralist*, and that his moralizing for children was clearly linked to his moralizing for doctors. The children who had visited the Bridgewater canal were now grown into professional men and home-staying ladies.

Percival's son James was a medical student at Edinburgh in the early 1790s; he died there in 1793. His father was to claim that *Medical Jurisprudence* was written for James' guidance; the bereavement almost caused the abandonment of the project. The eldest son, Thomas Basnett, had also intended a career in medicine, but he had returned from Edinburgh to Cambridge and entered the Anglican Ministry [18]. In 1791 he had preached a sermon in Liverpool in aid of the Infirmary; the text came to be included in his father's *Medical Ethics*. It stressed the educational/moral role of hospitals – the peculiar susceptibility to instruction there shown by the sick poor. (Perhaps donkey-drivers were more tractable when ill and confined in an infirmary.)

So one may easily argue that Percival was uniquely placed to turn an administrative request into an extension of public moralizing. But one may also venture another reason for Percival's self-assurance, for Manchester Infirmary was then no ordinary provincial hospital. It could claim to be one of Britain's most diverse medical charities, comprising a lunatic asylum and public baths, as well as a dispensary for out-patients and home patients. It offered medical education, and in spite of the new maternity hospital, it offered to deal with difficult labors. The poor, it was claimed, could there receive medical attention as good as any in the land. Percival, as its medical father-figure, may have felt a certain right to pontificate on medical services: 'that the official conduct and mutual intercourse of the Faculty might be regulated by precise and acknowledged principles of urbanity and rectitude' [41]. Here we note that Manchester, unlike Liverpool, had no formal Medical Society – it had the Infirmary and the Literary and Philosophical Society, both dominated by Percival.

But almost a decade passed from 1794 to 1803 when Percival finally published his modified work under the title *Medical Ethics*.

The dedication of Percival's text is to neither of the two sons already mentioned. By 1803, the parson son had died in St. Petersburg as chaplain to the British colony. This bereavement, too, threatened the father's literary output, but he rallied, in part because a third son, Edward, was then studying medicine. Edward was the dedicatee.

Percival's world was now darkening. He felt his age and the loss of his sons. His homiletic writings had become more serious and discursive, and not just because he now aimed them at adults. He engaged at length with theology, and worried about public religion [38]. His concern was also expressed in his revision of *Medical Ethics*, where doctors were enjoined to ensure they observed the Sabbath; public observance mattered for the *example* so set. After all, wrote Percival, there were so many kinds of churches and chapels, surely a doctor could find one which he can attend with good conscience [41]. It was a tolerant argument in a time of *decreasing* religious tolerance, for in Britain from the early 1790s, in the wake of the French Revolution and during the French Wars, politics were repressive and religion increasingly fearful and sectarian [62].

So, too, was medicine, at least if we are to judge from Manchester. The reform program with which Percival was associated lasted until after 1796, when the party which had reformed the Infirmary in 1790 managed to establish a Fever Hospital to counter the increasingly frightening epidemics of typhus. This Fever Hospital was the first such in Britain; it became a national model and was advertised by the Society for Bettering the Conditions of the Poor [48]. It was associated with a Board of Health – a voluntary civic venture well described by George Rosen [52]; the protagonist was John Ferriar, Percival's young colleague. One might reasonably regard the Board of Health and Fever Hospital as the zenith of a remarkably comprehensive set of medical charities which had grown from the reform of 1790. But by 1800 the reform policy was in decline, not least because of the crudest and most direct political repression (Thomas Walker, though once the leading citizen, was tried for treason; he was acquitted but ruined) [24]. The reformers had offered a general program of 'medical police', including municipal lodging houses for immigrant workers; by 1800 their institutions were defensive and conservative. New plans for expansion at the Infirmary were turned down; the Board of Health became little more than an agency responsible for the Fever Hospital.

And the Manchester medical profession became ever more contentious. There was now an institutionalized opposition between the Infirmary staff and the White group at the maternity charity. Professional issues were likely to polarize, rapidly and deeply, as did the 'Caesarean' dispute in 1799–1800. White and his 'school' of obstetricians included several early exponents of the Caesarean section, then a *very* rare operation. One such, John Hull, a surgeon-midwife turned Leiden physician, was newly established in Manchester. When he had the audacity to perform a Caesarean he was attacked as reckless by an Infirmary surgeon with an interest in midwifery. The resultant pamphlet war was truly vulgar, though decked out with long literary histories and classical quotations [19], [20], [22], [31], [54], [55], [57]. It was under such circumstances that the aging Percival revised and finally published his *Medical Ethics*, thirteen years after the Infirmary dispute which had first given rise to the proposal, and almost ten years after the first draft had been circulated. A year later he was dead.

So we return to the starting point, prepared to give a fuller answer to the question of why Percival's text was produced in the Manchester of 1790–1803. The argument I tentatively advanced, accords with that which I mentioned for Gregory. Percival's text, I would like to suggest, was written to defend, and so perhaps preserve, a passing order. Gregory did not write for radical anti-professionals, though some such may later have used his work. Percival did not write rules for regulating formal associations of professionals, though such associations may later have found his work useful. Percival's own project in Manchester was probably moribund by the time he published *Medical Ethics*. What, then, was this project?

We go back to the suburban villa and the young father instructing his children; to the neo-classical accord of nature, society and morals; to the Unitarian chapel of Thomas Barnes which Percival attended, where religion was more classical than hebraic; to the magistracy of Percival's friend, T. B. Bayley, a patrician in civic regulation, a major reformer and rebuilder of the local prison. We go to the Literary and Philosophical Society, to rational entertainment for a local intellectual élite. For a little while, in the 1770s and 1780s, Manchester may have seemed a plausible new Athens. It was relatively free of lordly influence; much of its civic business was transacted through voluntary societies; its dissenters may have been marginal to some aspects of English life, but they were powerful in Manchester, their chapel at least as rich as the local parish church. Several of the merchants and their sons were well-educated; Edinburgh graduates were available as conscientious magistrates and physicians; the town was prosperous, its suburbs expanding, its élite groups then in reasonable accord.

But Percival lived on past 1790 and into the repression and economic crises of the French wars. By the 1790s Manchester was experiencing the problems of rapid urbanization and industrialization [58]. The reform party responded creatively through a variety of voluntary charity activities which, in a sense, outflanked the rudimentary statutory agencies. So, men of culture and social enquiry became the formers or re-formers of a major set of social agencies; they were not just commentators on the urban poor, they were actors in local medico-civic action. Their programs should not be described merely as middle-class politics, for Percival did not represent all doctors any more than J. L. Phillips, treasurer of the Infirmary, represented all manufacturers: these men composed a self-conscious élite, with a culture they described as based on urbanity and rectitude, on the love of virtue, on a thirst for knowledge and an early acquaintance with the works of nature. Until it was suppressed or turned conservative, the Whig élite carried through a remarkable program of medical reform. They proved their influence and their capacity for public action. Robert Owen, then a local factory manager, was correct to describe them as a local aristocracy, if that meant a cultured governing group [56].

But such a system was frail. It would be crushed by the continuation of the challenges which had provided its scope for action, and by political repression in a nation fearful of revolutionary change. When circumstances finally improved, in the 1820s, some of the same families would emerge again in radical causes, but nineteenth-century liberalism was not eighteenth-century Whiggism; nor was there any longer the same single set of institutions which had made late eighteenth-century Manchester a relatively simple arena for civic action. The much larger town had become divided, by locality, religious denomination, and by professional groupings. In such circumstances doctors would increasingly rely on formal organization, rather than on the rankings of individuals in a well-ordered, small-town society.

Percival, it is often said, was defending an old medical hierarchy; that is true but insufficient. One has to recognize that his professional hierarchy was but part of a more general social hierarchy, without which it meant nothing. Percival was telling his "bretheren" how to behave if they were to preserve that system. Contention was its enemy. So, as a medical father-figure, he wrote a set of rules.

But why do I argue that Percival's was a *belated* defence of civico-medical virtue? Let me give two reasons: one general, one particular and empirical.

There is, I suspect, a general problem about literary defenses of social order. For as long as such an order has deep authority, books are unnecessary; once the authority is waning, books may be futile. A book on how to be a virtuous physician probably meant that the task was becoming more difficult and more problematical. Any such texts were unlikely to be guides to a new order; rather, they were guides to an old order which had *become* a problem.

The specific reason concerns Manchester and can be illustrated from yet another dispute: at the Infirmary, in 1801, the medical committee was up in arms about the powers of the lay governors. Such disputes over medical versus lay control became very common in the new century, and were a major feature of hospital politics, certainly into the 1860s. But we are rather well informed about the early Manchester case, not least from the annotated pamphlets left by the hospital treasurer, J. L. Phillips, whom I mentioned before; he was a key member of a Unitarian Whig family, grown somewhat conservative during the Wars. He must have been a long-standing acquaintance of Percival, probably a friend. Like Percival, he was by 1801 an aging representative of the Enlightenment, now living in harder times. What matters here is his attitude to the Infirmary doctors as they argued in 1801 about control of the House Surgeons, etc. The honorary staff were, still in a sense, men of charity, formally much the same as Percival. But such was not the view of Phillips, the mill owner. For him, these new doctors were on a par with the *managers* of mills – tools of the ruling élite, no longer partners [45], [63].

As we noted earlier, such doctors, from about 1830, would organize themselves into local and national medical societies to seek a new form of status and authority. Collectively, they would bid for civic recognition and for national representation [29]. In such bids, as indeed Waddington has shown, Percivalian codes had their uses – but mostly the politics were more direct [59]. You argued for reform of the Royal College of Surgeons or whatever; you proposed Acts of Parliament; you tried to ostracize homeopathy. Local medical societies, which became common in the 1830s, mostly concentrated on clinical and scientific reporting: Manchester's was founded around a Book Club [63]. Percivalian codes could provide decoration or distraction, but they were not going to recreate an élite of civic virtue in which cultured physicians could take their part. Percival himself could be recalled to evoke a generation of medical virtue, between the foolishness of aristocratic foppery and the overcrowded, factionalized medical world which had developed since the beginning of the wars [42]. But in their new societies, doctors were but doctors, whereas Percival and his friends had been full members of a single, recognized local élite. That situation had gone and could not be recreated, at least in such a city as Manchester had become.

We have seen that in a rapidly changing town enlightened reforms had proved contentious. The Infirmary had been expanded but controversy had flowed from the attempted rectification of medical hierarchy. Rules could be provided as remedy, but accustomed ways and deference were fading, and rules could not reverse the shift. Manchester's neo-classical culture had provided Percival's strength; the demands of new industrial and urban problems had created his opportunities for social action; the continuance and extension of these social and political changes came to render his own project irrelevant, except when new generations of medical reformers tried to invoke a lost authority.

Percival, I conclude, is not to be understood as if he were a nineteenth-century medical reformer. The conditions of production of his medical ethics were much more singular and peculiar. But, as I hope I have also shown, these conditions were of very general importance. In the dynamics of a single town over a single generation, one can explore changes of social structure and authority which are central to modern society. Such questions encompass medicine, but extend far beyond it.

John V. Pickstone
Wellcome Institute for the History of Medicine
University of Manchester

NOTE

* Stella Butler and I, in 1984, published a detailed analysis of the dispute around 1790 at the Manchester Infirmary in which the established hierarchy of Manchester medicine was violently disrupted [46]. This work has been carried on, not least in a thesis by Katherine Webb which contains detailed prospography of the local profession, on which I will draw [63].

REFERENCES

1. Aikin, J.: 1795, *A Description of the Country for Thirty to Forty Miles Round Manchester*, London.
2. Aston, J.: 1816, *A Picture of Manchester*, Manchester.
3. Baker, T.: 1884, *Memorials of a Dissenting Chapel*, London.
4. Berlant, J. L.: 1975, *Profession and Monopoly: A Study of Medicine in the United States and Britain*, University of California Press, Berkeley.
5. Bride, J. W.: 1922, *A Short History of the St. Mary's Hospitals, Manchester, 1790–1922*, Sherratt and Hughes, Manchester.
6. Brockbank, E. M.: 1904, *Sketches of the Lives and Works of the Honorary Medical Staff of the Manchester Infirmary from its Foundation in 1752 to 1830*, Manchester University Press, Manchester.
7. Brockbank, W.: 1952, *Portrait of a Hospital*, 1752–1948, Heinemann, London.

8. Buer, M. C.: 1926 (1968), *Health, Wealth and Population in the Early Days of the Industrial Revolution*, Routledge, London.
9. Clerke, W.: 1790, *Thoughts upon the Means of Preserving the Health of the Poor by Prevention and Suppression of Epidemic Fevers*, J. Johnson, London.
10. Davidoff, D. and Hall, C.: 1987, *Family Fortunes: Men and Women of the English Middle Class, 1780–1850*, Hutchinson, London.
11. Edelstein, L.: 1967, *Ancient Medicine*, ed. O. Temkin and C. L. Temkin, Johns Hopkins University Press, Baltimore.
12. Farrar, W. V., Farrar, K. and Scott, E. L.: 1973, 'The Henrys of Manchester, Part 1: Thomas Henry (1734–1816)', *Ambix* 20, 183–208.
13. Ferriar, J.: 1792–8, *Medical Histories and Reflections*, 3 vols., Cadell, London. [Printed in Warrington.]
14. Fulton, J.: 1933, 'The Warrington Academy (1757–1786) and its Influence upon Medicine and Science', *Bulletin of the Institute of the History of Medicine* 1, 50–80.
15. Gatrell, V. A. C.: 1972, 'The Commercial Middle-Class in Manchester, c.1820–57', Ph.D. Dissertation, Cambridge University.
16. Handforth, P.: 1956, 'Manchester Radical Politics, 1789– 1794', *Transactions of the Lancashire and Cheshire Antiquarian Society* 66, 87–106.
17. Hindle, G. B.: 1975, *Provision for the Relief of the Poor in Manchester*, 1754–1826, Manchester University Press, Manchester.
18. Hope, R. B.: 1947, 'Thomas Percival: A Medical Pioneer and Social Reformer, 1740–1804', M.A. thesis, University of Manchester.
19. Hull, J.: 1798, *A Defence of the Caesarean Operation*, R. & W. Dean, Manchester.
20. Hull, J.: n.d. [1799], *Observations on Mr Simmons Detection*, R. & W. Dean, Manchester.
21. Hunt, E. M.: 1959, 'The North of England Agitation for the Abolition of the Slave Trade, 1780–1800', M.A. thesis, University of Manchester.
22. Keown, J.: 1988, *Abortion, Doctors and the Law. Some Aspects of the Legal Regulation of Abortion in England from 1803 to 1982*, Cambridge University Press, Cambridge.
23. Kett, J. F.: 1964, 'Provincial Medical Practice in England, 1730–1815', *Journal of the History of Medicine* 19, 17–29.
24. Knight, F.: 1957, *The Strange Case of Thomas Walker*, Lawrence and Wishart, London.
25. De Lacy, M.: 1986, *Prison Reform in Lancashire, 1700–1850*, Manchester University Press, printed for the Chetham Society, Manchester.
26. Laqueur, T. W.: 1976, *Religion and Respectability: Sunday Schools and English Working Class Culture, 1780–1850*, Yale University Press, New Haven.
27. Lawrence, C. J.: 1984, 'Medicine as Culture: Edinburgh and the Scottish Enlightenment', Ph.D. Dissertation, University of London.
28. Leake, C. D. (ed.): 1927, *Percival's Medical Ethics*, Williams and Wilkins, Baltimore.
29. Loudon, I: 1986, *Medical Care and the General Practitioner 1750–1850*, Oxford University Press, Oxford.

30. McLachlan, H.: 1934, *The Unitarian Movement in the Religious Life of England*, George Allen & Unwin, London.
31. McLaren, A.: 1984, *Reproductive Rituals*, Methuen, London.
32. Meiklejohn, A.: 1959, 'Outbreak of Fever in Radcliffe Cotton Mills', *British Journal of Industrial Medicine* 16, 68–69.
33. Pellegrino, E. D.: 1986, 'Percival's Medical Ethics. The Moral Philosophy of an 18th-Century English Gentleman', *Archives of Internal Medicine* 146, 2265–69.
34. Penelhum, T.: 1985, *Butler*, Routledge and Kegan Paul, London, Boston and Henley.
35. Percival, E. C.: 1807, *The Works, Literary, Moral and Medical of Thomas Percival to which are prefixed Memoirs of his Life and Writings and a Selection of his Literary Correspondence*, 4 vols., new ed., J. Johnson, London.
36. Percival, T.: 1775, *A Father's Instructions to His Children; Consisting of Moral Tales, Fables, and Reflections Adapted to Different Periods of Life, from Youth to Maturity, and Designed to Promote the Love of Virtue, A Taste for Knowledge, and an Early Acquaintance with the Works of Nature*, J. Johnson, London.
37. Percival, T.: 1781, *A Socratic Discourse on Truth and Faithfulness Being the Sequel to a Father's Instructions*, Warrington.
38. Percival, T.: 1789, *Moral and Literary Dissertations Chiefly Intended as the Sequel to A Father's Instructions*, 2nd ed., revised and much enlarged, J. Johnson, London.
39. Percival, T.: 1794, *Medical Jurisprudence or a Code of Ethics and Institutes Adapted to the Professions of Physic and Surgery; with an Appendix containing a Discourse, addressed to the Gentlemen of the Faculty; the Officers; the Clergy; and the Trustees of the Infirmary at Liverpool, on their respective Hospital Duties*, privately circulated, Manchester.
40. Percival, T.: 1802, *Biographical Memoirs of Thomas Butterworth Bayley*, Manchester.
41. Percival, T.: 1803, *Medical Ethics*, J. Johnson, Manchester.
42. Percival, T.: 1827, *Medical Ethics*, W. Jackson, Manchester.
43. Percival, T.: 1849, *Medical Ethics*, J. H. Parker, Oxford.
44. Pickstone, J. V.: 1984, 'Ferriar's Fever to Kay's Cholera: Disease and Social Structure in Cottonopolis', *History of Science* 22, 401–419.
45. Pickstone, J. V.: 1985, *Medicine and Industrial Society. A History of Hospital Development in Manchester and its Region, 1752–1946*, Manchester University Press, Manchester.
46. Pickstone, J. V. and Butler, S. V. F.: 1984, 'The Politics of Medicine in Manchester, 1788–1792: Hospital Reform and Public Health Services in the Early Industrial City', *Medical History* 28, 227–249.
47. Porter, R.: 1980, 'Science, Provincial Culture and Public Opinion in Enlightenment England', *The British Journal for Eighteenth Century Studies* 3, 20–46.
48. Povey, W. P.: 1982, 'Aspects of Public Health in Manchester and District in the Late Eighteenth and Early Nineteenth Centuries. The Rise and Decline of the Manchester Board of Health 1796–1815', M.Sc. thesis, U.M.I.S.T.
49. Raymond, J. and Pickstone, J. V.: 1986, 'The Natural Sciences and the Learning of the English Unitarians. An Exploration of the roles of the Manchester

College', in Smith, B. (ed.): 1986, *Truth, Liberty, Religion: Essays Celebrating Two Hundred Years of Manchester College*, Manchester College, Oxford.
50. Redford, A.: 1939, *The History of Local Government in Manchester*, Vol. 1, Manchester.
51. Renaud, F.: 1885, *A Short History of the 'House of Recovery' or Fever Hospital in Manchester*, Manchester.
52. Rosen, G.: 1942, 'John Ferriar's 'Advice to the Poor', *Bulletin of the History of Medicine* 11, 222–227.
53. Rosen, G.: 1958, *A History of Public Health*, MD Publications, New York.
54. Simmons, W.: 1798, *Reflections on the Propriety of Performing the Caesarean Operation*, London.
55. Simmons, W.: 1798, *A Detection of the Fallacy of Dr Hull's Defence of the Caesarean Operation*, Manchester.
56. Thackray, A.: 1974, 'Natural Knowledge in Cultural Context: the Manchester Model', *American Historical Review* 79, 672–709.
57. Tomlinson, G.: 1799, 'A Letter to Mr. Ogden, Surgeon in Ashton-under-Lyne ... relative to the case of Elizabeth Thompson upon whom the Caesarean Operation was Lately Performed', printed by W. Shelmerdine & Co., Manchester.
58. Vigier, F.: 1970, *Change and Apathy. Liverpool and Manchester during the Industrial Revolution*, MIT Press, Cambridge, Mass.
59. Waddington, I.: 1984, *The Medical Profession in the Industrial Revolution*, Gill and Macmillan, Dublin.
60. Waddington, I.: 1975, 'The Development of Medical Ethics: A Sociological Analysis', *Medical History* 19, 36–51.
61. Walker, J. E. M.: 1973, 'John Ferriar of Manchester, MD: his life and work', M.Sc. thesis, Manchester, U.M.I.S.T.
62. Ward, W. R.: 1972, *Religion and Society in England* 1790–1850, Batsford, London.
63. Webb, K. A.: 1988, 'The Development of the Medical Profession in Manchester 1750–1860', Ph.D. thesis, Manchester, U.M.I.S.T.
64. Webster, C.: 1978, 'The Crisis of the Hospitals During the Industrial Revolution' in Forbes, E. G. (ed.): 1978, *Human Implications of Scientific Advance*, Edinburgh University Press, Edinburgh.
65. White, C.: 1773, *A Treatise on the Management of Pregnant and Lying-in Women*, London.
66. Young, J. H.: 1964, *St Mary's Hospitals Manchester, 1790–1963*, Livingstone, Edinburgh.

CHAPTER 8

ROBERT BAKER

DECIPHERING PERCIVAL'S CODE

Thomas Percival died in 1804, the year after the publication of *Medical Ethics,*[1] leaving nineteenth-century physicians free to appropriate his words without fear of contradiction from their author. The appropriation process began just four years after Percival's death, when the Boston medical society used his language to draft their medical police of 1808.[2] As American municipal, county, state and national medical organizations organized themselves from 1808 to 1846,[3] they followed the Boston precedent of prefacing their charters with codes of medical police or ethics, borrowing most of their language from Percival.[4] The process culminated in 1846 with the founding of what was to be the first national medical society, the American Medical Association (AMA). As the AMA still acknowledges, its original code of ethics drew heavily on Percival's words.

> At the first official meeting of the American Medical Association at Philadelphia in 1847, the two principle items of business were the establishment of a code of ethics and the creation of minimum requirement for medical education and training .... [I]t is clear that the AMA's first code is based on Percival's Code.
>
> In general the language and concepts of the original Code adopted by the Association in 1847 remained the same throughout the years. [Although] there were revisions ... in 1903, 1912, and 1947.
>
> ... The format of the Principles adopted in 1957 is a change from the format of the Principles promulgated by Percival in 1803, and accepted by the Association in 1847[1].

It is thus the official view of the AMA that, for its first century, its medical ethics was essentially identical with Percival's.

The identification of Thomas Percival with the nineteenth-century codes that appropriated his language is so complete that Percival himself has been appropriated as a nineteenth century figure – even though his dates, 1740–1804, place him squarely in the eighteenth century. So do his letters – among his correspondents were Diderot, D'Alembert, Franklin, and Voltaire – and most of his writings, including the work most highly

R. Baker, Dorothy Porter and Roy Porter (eds), The Codification of Medical Morality, 179–211.

regarded by his contemporaries, *A Father's Instructions Adapted to Different Periods of Life* (1st edition, 1775, 9th edition, 1803). Even the dates of the work which was ultimately to become *Medical Ethics* are eighteenth century. Percival commenced work on the project in 1792, when the trustees of the Manchester Infirmary asked him to draft a code of ethics to resolve a dispute that imperiled the institution. In 1794, he circulated a printed version of his code under the title *Medical Jurisprudence* to twenty-five eminent people (15 physicians, 4 barrister/lawyers, 3 clergymen/theologians, and 3 laypersons), whose comments formed the basis for a revised edition, published, in 1803, as *Medical Ethics*.

As the title of this essay indicates, what is attempted here is to disentangle the thought of Thomas Percival, the eighteenth-century physician and Enlightenment moral reformer, from the nineteenth-century codification project to which his words lent themselves. Other scholars to whom I am indebted have engaged in a similar project, most notably Chester Burns [4] and Edmund Pellegrino [14].

I. INTERPRETATIONS OF *MEDICAL ETHICS*

Although Percival's code was revered as the paradigm of professional medical morality in the nineteenth century, in the "muckraking" 1920s a revisionist interpretation reduced the standing of *Medical Ethics* to that of a mere professional etiquette. The originator of this interpretation was Chauncey Leake, a professor of pharmacology, and the editor of the 1927 edition of *Medical Ethics*. Leake was preoccupied with the lack of moral standards in contemporary American medicine. He believed this lack was, in large measure, attributable to the AMA's abdication of its responsibilities, which, in turn, was reflected in its code of "ethics." Hence he prefaced his edition of Percival with the following quote from Edward East:

> Since it is a kind of sacred writing to him, the young doctor doesn't appreciate the humor in terming a code of ethics that little guide to propriety issued by the American Medical Association, filled with trade union rules designed to promote dignity and prosperity in the profession ([13], p. vii).

It would be natural to assume that Leake wished to contrast the "trade union rules" which the AMA passed off as "ethics" with the genuine article penned by Percival. Leake's intention, however, was just the opposite. He pioneered the tradition of fingering Percival as the villain, or rather, the progenitor, of the AMA's confusion of ethics with trade union rules. To use the now classic canard, he accused Percival of originating the conflation of "ethics" with "etiquette." To prove this point, he appended the AMA codes of 1847, 1903, and 1912 to his edition of *Medical Ethics*, and provided a comprehensive cross-index, so that the reader could compare codes and verify Percivalian origins of the AMA's erroneous ways.

Leake's accusation – his location of the source of a current malady in a historic semantic confusion – was standard intellectual practice in the 1920s, and on this point, it is best to quote him directly.

> The term 'medical ethics', introduced by Percival is really a misnomer .... [I]t refers chiefly to the rules of etiquette developed in the profession to regulate professional contacts of its members with each other. Towards this, society often evinces a respect not warranted by considerations of it own welfare. Unfortunately Percival was persuaded that "medical ethics" was the proper title for his system of professional regulations. All similar and subsequent systems of general professional advice, whether official or not, have received the same title. As a result confusion has developed in the minds of many physicians between what may be really a matter of ethics and what may be concerned with etiquette ([13], p. 2).

At the heart of Leake's critique is the distinction between genuine ethics and etiquette. He characterizes genuine ethics as (i) "concerned with ultimate consequences of the conduct of physicians toward their individual patients and (ii) toward society as a whole"; it should also include (iii) "a consideration of the will and motive behind this conduct" ([13], p. 2); and (iv) should be predicated upon "analyses of ethical theory made by recognized ethical scholars" ([13], p. 3). Genuine medical ethics is thus grounded in ethical theory, and deals with the physician-patient, physician-societal relationships. On this characterization, a code of intra-practitioner relationships issued independently of recognized ethical theory is not a genuine ethic, it is only an etiquette.

Leake's presentation of Percival lay dormant until 1975 when two different writers, Jeffrey Berlant [2] and Ian Waddington [25] drew upon it to develop a social science critique of contemporary views of medicine. Waddington's critique, "The Development of Medical Ethics: A Sociological Analysis," appeared in *Medical History*; it targeted standard sociological theories which endorsed the view "that practitioner-client relationships are crucial to an understanding of professional ethics [which] has become almost a sociological orthodoxy" ([25], p. 36).

> Characteristically, those who pursue this line of argument suggest that for a variety of reasons, but primarily because of his ignorance, the client is unable to judge the quality of the professional service which he receives. Consequently, the client is very vulnerable to exploitation by the unscrupulous practitioner. The development of professional ethics is seen as a response to this problem of social control. Thus the professional group itself undertakes to guarantee the integrity of its members by the development and enforcement of codes of professional ethics ([25], p. 36).

Waddington argued that the orthodox sociological account "has been developed in the absence of any detailed empirical investigation of the development of codes of professional ethics" ([25], p. 38). When reexamined historically, it turns out that "the conventional explanation of codes is not valid ... even for the prototype profession [i.e., medicine]" ([25], p. 49). For "it was Percival who, more than any other person, effected 'the transition from the broad principles of Greek medical ethics to the current complicated system'" ([25], p. 36). Yet if Percival is really "the founder of modern codes of medical ethics ... [as] most modern men [believe]" ([25], p. 36), orthodox sociological theory of professional ethics must be in error because, "as has clearly been pointed out by Leake," *Medical Ethics* "is concerned primarily with regulating the relationship between practitioners ... [i.e] medical etiquette" ([25], p. 39).

Moreover, insofar as Percival actually does touch on medical ethics, he merely iterates the Hippocratic tradition.

> [Percival's] advice to practitioners on how to behave towards patients is, for the most part, of a highly general kind, very much in keeping with the Greek tradition; there is thus nothing specifically modern about it ([25], p. 38).

Thus, historically considered, the emergence of modern medical ethics is just the opposite of what orthodox sociological theory would lead one to expect: it is *not* a revision in physician-patient relationships, but a reconceptualization of intra-practitioner relationships. What passes for professional ethics, is really intra-practitioner etiquette.

Moreover, this etiquette of intra-practitioner relationships is not concerned with the protection of the patient, but is an attempt by practitioners to restrict competition for patients – a problem exacerbated by the tripartite division of British medicine into apothecaries, physicians, and surgeons. Percival's solution to the problem of competition is essentially "conservative," designed merely to "prevent the continual disputes over the division of labour within the profession ... [while] ...

maintain[ing] ... the traditional divisions within the profession"([25], p. 44). In the end, his solution was ineffective since the problem "smoldered on throughout the nineteenth century" ([25], p. 47). In Waddington's view, the fecklessnes of medical ethics is characteristic of the ethical enterprise in general.

Medical practioners, it is clear, were no more given than any other section of the educated classes to the consideration of abstract philosophical principles. Rather their concern with medical ethics was a practical concern, arising from certain recurrent problems with which they were faced in the day to day practice of their profession ([25], p. 48).

A remarkably similar line of argument was offered in the same year by Jeffrey Berlant in *Profession and Monopoly: A Study of Medicine in the United States and Great Britain* [2]. Like Waddington, Berlant portrays Percival's code as conservative and argues further that the American codifiers were drawn to Percival because his "medical ethics [was] ... a means of legitimatizing the monopolistic privileges of the profession to the powers that be and to the public."

The conservative Percival wrote his ethics at a time when the elitist medical corporations had come under democratic attack, particularly by economic liberals. As an apology for the corporations, Percival's work was superb; as a means for integrating the profession, it was unsurpassed. As an instrument for solidarity it was particularly suitable for extending the monopolistic controls of the RCP [Royal College of Physicians] to the newly professionalized surgeons and apothecaries. Therefore, on the basis of this comparative study, one can say that Percivilean ethics were probably the organizational tool for bolstering the system of licensing corporations through the encouragement of monopolistic traditions for all professions, an important device for suppressing competition between different types of professions which might have been exacerbated by the appearance of increasing numbers of professional corporations ([2], p. 59).

Unlike other revisionists, however, Berlant holds that Percival fashioned a medical ethic, albeit an unenforceable one, on the ethic of *virtue*. Percival presupposed

... the fundamental goodness of human nature and on this basis rejects the need for sanctions .... Percival evidently believed in the power of revealed reason: Physicians have a single most rational way to behave and once shown the way, will follow. Percival's role is to be the voice of reason.

As a social philosopher, Percival stands as a naively saintly man: he disdains conflict, bourgeois competitiveness and group coercion; he envisions conscientious men, professional identification with a higher cause, and the superiority of ethical wisdom over social control. Percival embraces the ideal of a strong profession, composed of strong members committed to the goodness of their cause; how this might be accomplished did not, as with many Enlightenment thinkers, seem problematic to him. Presumably the rationality of men once shown the way would be sufficient ([2], p. 56).

II. THE REVISIONIST READING OF *MEDICAL ETHICS* RECONSIDERED

The revisionist reading of Percival rests on four assertions: (1) that the subject of *Medical Ethics* (and the subsequent AMA codes) is primarily the regulation of intra-practitioner relationships, not the practitioner-patient relationship; (2) that these regulations are properly considered an etiquette, not an ethic; and (3) that their intent is essentially conservative, i.e., to preserve the monopolistic powers of the profession. This reading is seemingly confirmed by (4) the apparently incontrovertible historical fact that the American codifiers, by simply copying Percival, were able to set the organizational framework for a strong profession which aimed at monopolistic control of medical practice.

Point (1) of the revisionist reading, has been definitively rebutted by Chester Burns [4] and Edmund Pellegrino [14]; fewer words, however, have been directed at point (2) – especially Leake's original contention that Percival himself unknowingly conflated etiquette with ethics. Curiously, Leake adduced no textual evidence in support of this contention. In light of Percival's acknowledgement of his debt to John Gregory *Lectures* [7], it would be odd to find Percival unaware of this distinction, or inadvertently conflating it, since Gregory's second lecture deals with the moral irrelevance of decorum and etiquette. Turning to the text of *Medical Ethics*, we find Percival taking cognizance of the ethics-etiquette distinction and treating them as distinct. Consider, as a case in point, Percival's remarks in Chapter Three, Articles III and IV. In III, he observes that:

> This amicable *intercourse* and *co-operation* of the physician and apothecary, if conducted with the *decorum* and attention to *etiquette*, which should always be observed by professional men, will add to the authority of the one, the respectability of the other, and to the usefulness of both" (p. 55, italics in the original).

In this passage, Percival specifically characterizes intra-practitioner cooperation in terms of "*decorum*" and "*etiquette*" while eschewing the language he uses elsewhere to characterize "*moral rules of conduct* ... towards hospital patients" (Chapter Two, Article I), i.e., the language of duty and responsibility.

The moral language of rules of conduct surfaces, however, immediately afterwards in Article IV:

The duty and responsibility of the physician, however, are so intimately connected with these points, that no dependence on the probity of the apothecary should prevent the occasional inspection of the drugs which he prescribes" (Chapter Three, Article IV, p. 55).

Percival is here asserting that the physician's moral "duty and responsibility" to inspect drugs (i.e., to protect the patient), *overrides* the *etiquette* of cooperation with the apothecary described in Article Three. A careful reading of the text thus indicates that Percival not only distinguishes between ethics and etiquette, he explicitly holds that the ethical duties override the *etiquette* of cooperation.

Although textual analysis invalidates the charge that Percival inadvertently conflated the concepts of ethics and etiquette, the gist of the Leake and Waddington critiques can be preserved if one argues that what is involved is a *misidentification*, rather than a conflation; i.e., that Percival (and the AMA), in point of fact, misidentify an etiquette as an ethic. Leake, in effect, argues this point by specifying four essential characteristics of a medical ethic and then arguing that the codes fashioned by Percival (and the AMA) lack all four of these characteristics. Burns and Pellegrino have successfully demonstrated, however, that the text of *Medical Ethics* easily satisfies all four of Leake's criteria.

An alternative reply to the revisionists is to challenge the criteria Leake uses to characterize an ethic. Consider (iii), which states that an ethic must involve "consideration of will and motive behind conduct." Were this a necessary condition for being "ethical" it would deny the status of "ethics" to the works of Percival's contemporary, Jeremy Bentham – the founder of utilitarianism. Criterion (ii), is equally problematic; it requires a genuine professional ethics to deal with the relationship between the profession and society. Both the Hippocratic Oath and the lectures of John Gregory are silent on this subject. Are these paradigms of medical ethics to be dismissed as etiquettes as well? Or, are the criteria used to reduce Percival's *Medical Ethics* to an etiquette excessively stringent?

It would seem that the issue is not whether *Medical Ethics* satisfies revisionist criteria for a "real" ethic, but why one should accept the revisionist characterization of "medical ethics" as definitive. Is it reasonable to accept any definition of "medical ethics" which excludes such paradigm cases of medical ethics as Percival's *Medical Ethics* and the 1847 AMA code? Leake feels justified in taking this extreme step because he believes that the codes written by Percival and the AMA are not predicated upon "ethical theory" as "recognized by ethical scholars."

In advancing this argument, Leake is evoking what is sometimes characterized as an "applied ethics model," i.e., he presupposes that ethics is the application of formalized (philosophical) moral theory to specific cases. In my view, such models mistakenly invert the relationship between the phenomenon being theorized about, and the theory. The characterization of conduct as ethical is a mundane cultural activity which people engage in quite independently of any scholarly theory of ethics. Ethics, in this sense, is like economics – a cultural artifact essential to the functioning of complex societies, which exists separately from the scholarly theories that describe, critique, rationalize, and attempt to modify it. And just as economic activity can exist in the absence of economic theory, so too can ethics.

Revisionists contend that Percival and his contemporaries, acting without illumination from ethical theory, somehow mistook etiquette for ethics. Is this claim reasonable? Or, is it more reasonable to suspect that twentieth-century readers of eighteenth-century texts, confused the ethic of another era with an etiquette? Historians have a term for such confusions – "presentism" – the fallacy of reading the past as if it were the present. Isn't that the most plausible interpretation of the reading offered by the two non-historians, Leake and Waddington? They were motivated by contemporary controversies, and they read Percival in terms of these controversies; in so doing, they implicitly demanded that an eighteenth-century moralist use the language and emphasize the issues considered most relevant in the twentieth century.

Presentist conceptions of medical politics also appear to underlie the "conservative" label Berlant and Waddington attempt to pin on Percival. They justify this label by casting Percival in the role of "apologist" for the Royal College of Physicians who wished to "extend their monopolistic traditions for all professions." Yet Percival was an Edinburgh-Leiden educated Dissenter and an Enlightenment moral and sanitary reformer who practiced medicine in Manchester, i.e., outside of London. He never was a member of the Royal College of Physicians (RCP), nor was he a partisan of this particular "medical establishment." The anonymous editor of the (2nd) 1827 edition of *Medical Ethics*, who is unstinting in his praise of Percival, opens the book by declaring his hostility to the RCP – a position he was unlikely to have taken, at least in the form he does, were Percival an apologist for the RCP. A short quote establishes the flavor of the anonymous editor's remarks.

> Concerning the Colleges of Physician in London and Dublin, though we would not be esteemed setters forth of strange doctrines, nor pullers down of strongholds, nothing shall here be offered in apology for the statements contained in this work. These superannuated institutions are millstones and dead weights hung around the profession, which, however well adapted to it three centuries since, are now little superior to mere vehicles of diploma mongering, and of perverted and illiberal views [1827, ed. xxiv].

Indeed, one of the reasons the editor despised the RCP is that it rejected Percival's application for membership, apparently because he was not English-educated: "Percival's name floats down the stream of time, with those of Fothergill, Dobson, Darwin, Currie, and others who will not be forgotten, although an English University did not contribute in any way to save them from oblivion" (xxiv). The revisionist reading of Percival as an apologist for the RCP, requires, in short, not only a revision of our perception of Percival, but of the established facts about his life and times.

Charges of "presentism" can also be leveled against Waddington's line of argument. He casts Percival as a conservative by comparing passages from Chapter Three of the *Medical Ethics* with mid-to-late *nineteenth-century* debates over dissolving the tripartite status/education hierarchy of the British medical guilds. Thus, Waddington condemns Percival as a "conservative" largely on the basis of the opening line in Chapter Three, Article I: "*In the present state of physic*, in this country, in which the profession is properly divided into three branches, a connection peculiarly intimate subsists between physician and the apothecary, and various obligations result from it" (emphasis added). Waddington ignores the first six italicized words in the line quoted, and then juxtaposes it with two statements about nineteenth-century issues: the first, about *The Lancet*, a journal which "since its foundation in 1823, campaigned consistently for the abolition of the tripartite structure," ([25], p. 46), and the second, an 1834 quotation published anonymously in the *British and Foreign Medico-Surgical Review*. It seems rather odd (although entirely in keeping with the presentist style of the revisionists) to castigate Percival for being "conservative" because he addresses contemporaneous issues, the practices of the 1790s (i.e., "the present state of physic"), rather than problems of the 1820s or the 1840s (i.e., issues which will occupy the limelight some twenty to forty-five years in the future). It is especially odd because both in his rhetoric and in the content of his rules Percival always treats the two medical specialties as equals and (unlike Gregory) addressed *Medical Ethics* to both physicians and surgeons.

For the revisionists, the ultimate proof of *Medical Ethics'* "conservativism" is really (4), that by simply copying Percival, the American codifiers erected the architectonic of a conservative monopolistic profession. (It is difficult to understate the importance of this point for Berlant and Leake.) The remarks prefacing the 1847 American code certainly seem to support the notion that the Americans simply copied Percival.

> The members of the convention ... [will] recognize in parts of [the code] expressions with which they were familiar. On examining a great number of codes of ethics adopted by different societies in the United States, it was found that they were all based on that by Dr. Percival, and that the phrases of this writer were preserved, to a considerable extent, in all of them. Believing that language so often examined and adopted, must possess the greatest of merits for such a document as the present, clearness and precision, and having no ambition for the honor of authorship, the Committee which prepared this code have followed a similar course, and have carefully preserved the words of Percival wherever they convey the precepts it is wished to inoculate ([13], p. 218).

Yet, despite this apparent admission that they simply copied Percival, a careful comparison of the American codes with the 1803 edition of *Medical Ethics*, suggests a more complex relationship. The codifiers did indeed preserve Percival's language, but they also tailored his text to fit the American context. American egalitarianism left them little choice. Percival was addressing a class and status-conscious culture in which distinctions between the upper and middle classes, the working classes and the poor, were evident and important to everyone – as was the status and specialization distinctions between university-educated physicians, hospital-trained surgeons, and apprenticeship-trained apothecaries. In Britain, *Medical Ethics* was received as a work of Enlightenment reform precisely because Percival tried to surmount these differences, arguing, for example, that the sick-poor in hospitals ought, insofar as possible, be given the same treatment accorded to more affluent classes (Chapter One, Article II).

In the more egalitarian medical and popular culture addressed by the American codifiers, however, these distinctions were not recognized.

Thus, where Percival addressed "physicians," "surgeons," and "apothecaries," the American codes excised these terms and replaced them with a single word, 'physician'. Again, Percival constantly refers to different categories of practice and patients: private, dispensary, infirmary, lock hospital, and insane asylum. The American codes, however, use just one term 'patient'. Ironically, the egalitarian excision of these different categories of patient also bowdlerized the Enlightenment reformism of Percival's ethics. For in expurgating un-American distinctions between classes of patients and practitioners, the codifiers also eliminated the contexts in which it made sense to exhort that all categories of patient be treated equally. They, thus, deleted all of Percival's detailed warnings about the dangers facing female infirmary patients (Chapter One, Article V), Lock Hospital patients (Article XXVII), and insane asylum patients (Articles XXVIII–XXXI; see also Chapter Four, Articles III–V).

The Americans excised other foreign elements from Percival as well. It was natural for Percival, a British subject, to conceptualize the privileges of medical practitioners as bestowed upon them by society acting through government. English medical practitioners had their privileges bestowed by Royal charters granted during the reign of Henry VIII to the Barber-Surgeons Guild (1512) and to RCP of London (1518). These notions were unnatural to the Americans, who, in the years 1808 to 1847, were in the process of creating professional organizations without government grants. The very idea that organizations require governmental recognition affronted the Lockean heritage of the Americans, who in the *Declaration of Independence*, *The Constitution*, and the *Bill of Rights*, envision governmental powers as deriving directly from the people, and the powers not explicitly delegated to the government, as reserved to the people themselves. These difference are reflected in most American codes, including the 1847 AMA code which, unlike Percival's, takes the form of an explicit tripartite social contract directly between professionals, patients, and the public. This Lockean conception of a direct contract with patients appears to have been inspired by Benjamin Rush's essay of 1794 – "Duties of a Physician"; it is perhaps worth remarking that Rush was, in fact, a signer of the *Declaration of Independence*.

The differences between American and Percivalean conceptions of the relationship between the profession and government become evident in those sections of the codes which deal with what we would today call "confidentiality." Percival argues in Chapter Four, Article XIX, that practitioners should not be misled by "false tenderness or misguided conscience" into "withholding the necessary proofs" when testifying in criminal court. On the contrary, the practitioner is obligated "not to conceal any part of what he knows, whether interrogated particularly to that point or not" (pp. 110–111). The AMA code, in contrast, replaces such discussions with sections on "secrecy" which explicitly state that: "The force and necessity of this obligation [to secrecy] are indeed so great, that professional men have ... been protected in their observance of secrecy by courts of justice" ([13], p. 220). Thus, where Percival conceptualized practitioner-patient obligations as properly controlled by government, the Lockean Americans see professional-patient obligations as a private matter, deriving from a direct tripartite contract betweens professionals, patients, and the public. The 1847 code is their version of this contract – which they seek to have recognized and protected by government (or at least "courts of justice").

Returning again to the problems of ahistoricism that plague the revisionist reading of *Medical Ethics,* in an important 1974 paper, "Reciprocity in the Development of Anglo-American Medical Ethics, 1765–1865," Chester Burns has worked out the chronology of nineteenth-century codification of medical morality. Unlike almost everyone else who has written on the subject, Burns' account is constrained by his reading of the historical record. Consequently, the tale he tells is more convoluted. The gist of the story is that the sequence presumed by most commentators inverts the order of certain crucial events. The standard presumption is of a Percivalean codification in Britain during the first half of the nineteenth century, which becomes the model for the American codification of 1847. Burns argues that although Percival, like Gregory, was an influential British writer on medical ethics, the British had no professional code of ethics, i.e., no code formally accepted as a code of ethics by any national medical organization until at least the late 1860s. In contrast, as early as 1808, American medical organizations developed codes drawn almost directly from Percival – a process which culminated in the development of a national code by the AMA in 1847.

Burns' chronology raises serious questions for revisionist readings. Waddington's analysis, for example, presupposes that Percival authored a "code of professional ethics," in the sense which "orthodox sociologists" use this expression. I.e., he presupposes a formal ethic, endorsed and maintained by a profession, characteristically through professional organizations (such as the AMA and the BMA), or, failing that, a *de facto* ethic, accepted by professionals, even without an organizational imprimatur, as the standard by which members of a profession are to be judged. Except in the Manchester Infirmary itself, however, Percival's code never functioned as a professional ethic (in this orthodox sociological sense) either in Britain or, despite its influence, in America. It would seem, therefore, that the code Waddington selected as the paradigm of a "professional ethic" in the orthodox sociological sense, never was a professional ethic in the narrow sense of the term – it was, at best, a proposal for such an ethic.

Waddington, it should be remarked, provides no evidence that Percival's code was accepted by any British medical organization, nor that it was accepted *de facto* by the British physicians of the period. What Waddington offers in lieu of this are statements by three *twentieth-century* physicians, Forbes, Barton, and McConaghey, who iterate the standard view that Percival "compiled the first modern code of ethics" ([25], p. 36). This is not the same thing. If we look at material from the nineteenth century, e.g., the anonymous 1827 edition of *Medical Ethics*, it would appear that Percival's was neither a *de jure* nor a *de facto* code of medical morality, since the purpose of that edition was to chide the British profession for its lack of moral standards. The anonymous editorship of the 1827 edition underlines the status of Percival's code as a code of ethics *offered* to, but never accepted by, the British medical profession. This corroborates Burns' thesis that the *first* modern professional code of medical ethics is not Percival's code, but either the Boston Medical Police of 1808 or, if such a code is deemed too parochial, i.e., too regionally limited to represent the ethic of an entire profession, the AMA code of 1847 (which, by fiat of the national organization, superseded all regional codes in 1855).

If the American national code of 1847 is the first modern English-language professional ethic, Waddington's case against the orthodox sociological account of the professions collapses, as does his entire account of the codification of medical morality in Britain and his explanation for it. For the American code of 1847 is self-evidently concerned with the practitioner-patient relationship in precisely the way predicted by orthodox sociological accounts. As the three chapter titles to the American code make evident, the profession's manifest intent in drafting the code is to *establish a contract*, a *quid pro quo, with the public*: the profession, on its part, pledges internal regulation and service in exchange for a societal ratification of the profession's autonomy and prerogatives. This is precisely the scenario predicted by conventional sociological theory, and so a careful historical study of the origin of professional codes of ethics would appear to validate the orthodox, rather than the revisionist account.

Moreover, Waddington's account of the impetus for codification can not be correct. On his view, codification is a conservative response to the problems of the disintegrating tripartite division of labour in Britain. Yet the American codification can have nothing whatsoever to do with the preservation or disintegration of the tripartite division of medical labour, since this division, while indigenous to Britain, was absent from egalitarian America.

Waddington's critique of Percival's code *per se* is, of course, independent of his account of the codification of medical morality. If it is true, as he contends, that Percival's (and/or the AMA's 1847) "advice to practitioners on how to behave towards patients is, for the most part, of a highly general kind, very much in keeping with the Greek tradition" then, indeed "there is ... nothing specifically modern about it" ([25], p. 38). Waddington's point here is that since the practitioner-patient relationship remains constant from Greek times to "the first modern code of ethics," what made Percival "modern" had nothing to do with a change in practitioner-patient morality and everything to do with a radical revision in the conception of intra-practitioner relationships.

A cursory look at Chapter One of *Medical Ethics*, however, will immediately establish that Waddington is wrong. Percival does not give behavioral advice of a highly general kind on the Greek paradigm; he states specific detailed duties towards hospital patients. Article II, for example, directs practitioners to accommodate the hospitalized patient's request to be treated by a "favorite practitioner;" Article IV directs practitioners not to discuss cases in front of ward patients; Article VIII directs practitioners to resist Trustee parsimony when it is deleterious to the patient's welfare; Article XII addresses the regulation of experimentation on human subjects. The list goes on, but the articles cited should suffice to establish not only that Percival dealt with specific duties, but with specific newly emergent "modern" problems, i.e., those raised by a new type of patient, the hospital patient.

The eighteenth-century charity hospital, which specialized exclusively in the care of the sick, was a social innovation, and the Manchester Infirmary was a preeminent example of it. Yet even this paragon was burdened with problems of overcrowding, underfunding, and was divided into wings demarcated according to newly developed distinctions between infirmary, lock hospital, recovery hospital, lunatic asylum and so forth. Percival's "modernity" thus arises almost as a matter of course. In setting moral standards for an new institution – an institution where, in effect, no standard existed previously – Percival of necessity broke new moral ground. No one before him, (not even Gregory) dealt with the practitioner-patient, intra-practitioner or practitioner-trustee relationships in a hospital context. Whatever Percival wrote, therefore, was necessarily modern medical morality, for he was the first to set ethical standards of behavior for that essentially modern institution, the hospital.

III. "DECIPHERING" PERCIVAL'S CODE

Percival's code is no cipher – except to those who insist on a revisionist reading. The revisionists' erroneous reading of *Medical Ethics* stems, in part, from the confusion of his authorship of the first modern *code* of medical ethics, with authorship of modern medical ethics *per se*. Percival lays no claim to being the inventor of modern medical ethics – an honor he properly accords to others. He claims only to be the first to present modern medical morality in the "aphoristic form of a code." He explicitly acknowledges drawing on four sources in developing this code. The first, and in many ways the most important, was the findings of an advisory committee of physicians and surgeons which, among other things, assembled rules of conduct used by hospitals and infirmaries in England and Scotland. The second and third explicitly acknowledged influences were "the excellent lectures of Dr. [John] Gregory" (p. 6), i.e., Gregory's 1772 *Lectures on the Duties and Qualifications of a Physician* and the various editions of Reverend Thomas Gisborne's 1794 *An Enquiry into the Duties of Men in the Higher and Middle Classes of Society in Great Britain Resulting from Their Respective Stations, Professions and Employment*. Percival's fourth source of inspiration was the comments of the twenty-five correspondents to whom *Medical Jurisprudence* had been sent. The correspondence with Gisborne appears to have been particularly helpful; the two authors publicly acknowledge borrowing each other's ideas ([7], p. 383; Percival, p. 6, 186–189), while at the same time criticizing aspects of each other's work ([7], pp. 401–403; Percival, Note VII, pp. 156–169).

Percival's achievement in *Medical Ethics* was thus brilliantly syncretic; the work fused modern (or, rather, eighteenth-century) moral theory to the practices of newly emergent medical institutions, thereby creating a synthesis of ethical theory and moral pragmatics that was to dominate Western medicine, especially in the English-speaking world, for over a century. The conceptual underpinnings of Percival's fusion were laid in Gregory's lectures, which reinterpreted "the art of medicine," in terms of eighteenth-century theories of moral sense. The architectonic which Percival used to structure his ethic was largely derived from Gisborne, while the content of most of the specific articles in Percival's codes was supplied by the researches and recommendations of the advisory committee.

The language of moral sense which infuses *Medical Ethics* derived less from Percival's earlier writings than from Gregory's *Lectures*. A perusal of Percival's use of moral sense terminology reveals that when he uses it to characterize the moral psychology of private practice medicine he borrows directly from Gregory. Chapter II, Article I is a case in point:

> I. The *moral rules of conduct*, prescribed towards hospital patients should be fully adopted in private or general practice. Every case, committed to the charge of a physician or surgeon, should be treated with attention, steadiness and humanity: Reasonable indulgences should be granted to the mental imbecilities and caprices of the sick: Secrecy and delicacy when required should be strictly observed.

The sentiments of the moral physician cataloged here – attention, steadiness, humanity, secrecy and delicacy – are essentially Gregory's.

When one turns to the counterpart passage in Chapter One, Article I, however, one finds Percival using different moral sense terminology.

> ... Hospital physicians and surgeons should study, also, in their deportment, so to unite *tenderness* with *steadiness*, and *condescension* with *authority*, as to inspire the minds of their patients with gratitude, respect, and confidence.

What accounts for the difference? Why does Percival deviate from Gregory's terminology and introduce the terms 'tenderness' and 'condescension' in Chapter One and then revert to Gregory's terminology, i.e., 'attention' and 'humanity', in Chapter Two? The evident reason appears to be that Chapter Two deals with the very same class of patients that Gregory was discussing in his lectures, i.e., *private patients*. The patients referred to in Chapter One, however, were the *sick-poor* of the community, hospital patients. Thus in Chapter One, Percival had to adapt Gregory's moral psychology to a context in which practitioners were treating hospital patients, who were from a lower social status, were not in any sense patrons, and whose opinion was unlikely to affect the practitioner's reputation with either his regular patients, his colleagues, or institutional trustees.

Strange as it may seem today, Percival's advocacy of "condescension" is crucial to his attempt to introduce Gregory's moral sense analysis of the psycho-dynamics of the physician-patient relation to the situation of hospital physicians and surgeons managing the care of charity patients. 'Condescension' is a term whose meaning has become inverted over the course of time. In our present egalitarian age, 'condescension' implies a pretense to (presumably undeserved) superior status. In the openly hierarchical status-conscience eighteenth century, however, condescension was a measure of temporary egalitarianism; 'to condescend' was to treat someone of lower status *as if* they were equals, suspending for the moment real differences of rank and status. Thus, the greater the distance in status, the greater the degree of condescension required to appreciate the patient's feelings and emotions, which is precisely the point Percival made to his middle class practitioners in Chapter One, Article XXVI: "greater *condescension* will be found requisite in domestic attendance on the poor" (p. 24). Read in the eighteenth-century context, Percival's urging of *condescension* and *tenderness* emerge as radically egalitarian attempts to secure for the sick poor the same sort of psychological relationships that Gregory had urged as morally requisite for sick private patients.

Another measure of Percival's egalitarianism is the rather bold first line of Chapter Two. There Percival states, "The *moral rules of conduct*, prescribed towards hospital patients should be fully adopted in private or general practice," i.e., the same rules of morality apply to all persons, rich or poor. Notice that Percival not only states this substantively, but stylistically, by establishing rules first for the poor (in Chapter One) and then applying them, *mutatis mutandis*, to middle class private patients in Chapter Two.

Many a moral reformer shared Percival's sentiments; few were bold enough to follow his style. Gisborne, for example, follows the more conventional approach by first stating moral obligations of physicians toward "patients in the upper and middle ranks of society" and then offhandedly remarking that "the greater number have likewise an obvious reference to his [i.e., the physicians] duty when visiting the poor" ([7], pp. 406–407).

Percival's egalitarianism appears to be predicated upon two virtues he takes to be fundamental: justice and humanity. Thus in Chapter One, Article II, Percival states that it is "just and humane to inquire into and indulge [the] partialities" of the "sick poor," in choosing physicians and surgeons, for they are as much in need of "self-confidence" and "comfort" as are "the rich under comparable circumstances." In Article I, he urges practitioners to hold themselves to a standard of "conduct" which will "inspire the minds" of their sick-poor patients "with gratitude, respect, and confidence," i.e., with the same mental states that practitioners traditionally seek to inspire in their middle class patrons. Despite, or perhaps, because of his commitment to egalitarianism, Percival appeals to practical and *prudential* interests of practitioners in his discussion of their conduct towards the sick-poor. He urges allowing the sick-poor a choice of practitioners on the prudential grounds that, when not so indulged, patients often "request their discharge, on a deceitful plea of having received relief, and afterwards procure another recommendation, that they may be admitted under the physician or surgeon of their choice" (p. 10).

Percival also argues that it is in the practitioner's self-interest to conduct themselves with "tenderness" towards their sick-poor patients. Like Gisborne and Gregory, Percival believed that medical practice tended to "pervert" the character of the clinician by engendering "coldness of heart" and decay of "tender charity" and that "This coldness of heart, this moral insensibility, should be sedulously counteracted before it has acquired an invincible ascendancy" (p. 180). Proper conduct, such as attending public worship, or acting tenderly towards patients, is seen an anodyne against "insensibility." By proper conduct, one forms *"active propensities"* that "gradually strengthened, by the like renewal of the circumstances which excite them," (i.e., by attending religious services, or by acting tenderly towards patients, the practitioner induces actual feelings of piety or tenderness in himself). Conduct, thus, cultivates virtue by which good "is steadily pursued, without struggle or perturbation." (For more on this point, and for the views of Gisborne and Gregory see the discussion of public worship in Note XII of *Medical Ethics*.)

Because he was writing an ethic for hospitals, Percival had to deal not only with the standard subject of the practitioner-patient relationship, but with a set of relationships largely ignored by Gregory and Gisborne, intra-practitioner relationships and practitioner-trustee relationships. These relations had to be rethought because the sick-poor patients treated in the Manchester Infirmary and other hospitals were not patrons. Indeed, as

Percival's comments make clear, he was acutely aware of potential conflicts between the interests of the hospital practitioner's patrons, the trustees, and those of their sick-poor patients. Thus, he indicates that trustees were inclined to overcrowd wards (Article XVI), to economize on effective medications (Article VIII), to proselytize patients (Article VI), and so forth. A complete code of conduct for hospital practitioners, therefore, would not only have to provide guidance on how practitioners should interact with each other, but also how they should act when the interests of hospital patients conflicted with those of their trustee/patrons.

Gregory's lectures provided little guidance on these matters. Gregory almost invariably portrays a physician engaged in solo private practice in competition with other practitioners. His argument for moral conduct, in fact, is partially based on the presumption that cultivating moral sensibility (i.e., feelings of humanity and sympathy towards one's patients) will "promote the private interest" of the physician ([8], p. 14) by providing him with the competitive advantage – for the scrupled physician can offer the patient sympathy, "an attention which money can never purchase" ([8], p. 22). Such market considerations had no persuasive power for hospital-based practitioners, since hospital patients were essentially a captive market. Consequently, if Percival was to offer hospital practitioners grounds for cultivating an attitude of moral sensibility towards their charges, he would need to look elsewhere.

Anyone conversant with English-language ethics at the turn of the eighteenth century, will realize, after reading the first words of Percival's code, that among the "elsewheres" Percival's looked, was Reverend Thomas Gisborne's *An Enquiry into the Duties of Men in the Higher and Middle Classes of Society in Great Britain Resulting from Their Respective Stations, Professions and Employment* (or perhaps, Gisborne's earlier work of 1789, *The Principles of Moral Philosophy Investigated and Briefly Applied To the Constitution of Civil Society*). In these works, Gisborne defends a social contractarian analysis of duty against William Paley's critique of John Locke. Gisborne's basic contention was that those who accept the protection of society's laws and/or the privilege of various offices or stations have *tacitly* contracted to accept a set of duties as well. In the *Enquiry*, Gisborne delineates the specific duties of individuals in accordance with their stations, or offices: magistrate, justice of the peace, lawyer, and, of course, physician.

With this in mind, it is possible to provide a Gisbornean reading for the first line of *Medical Ethics*.

I. HOSPITAL PHYSICIANS and SURGEONS should minister to the sick, with due impressions of the importance of their office; reflecting that the ease, the health, and the lives of those committed to their charge depend on their skill, attention, and fidelity.

'Office' is a term Gisborne used in arguing that those in various stations have entered into a tacit agreement with society. Such persons may properly enjoy the prerogatives of station only insofar as they satisfy the obligations of office. Percival can thus be read as stating that the office of hospital physician or surgeon confers a tacit (Gisbornean) obligation to conduct themselves towards their charges using skill, attention (one of Gregory's attitudes of moral sense), and fidelity, so as to promote the ease, health, and life of those committed to their care.

A similar Gisbornean note is struck in the first article of Chapter Four, where, in language reminiscent of the first article of Chapter One, Percival discusses the reciprocal relationship between government's conferral of privileges on the profession, and its expectations that the profession will assist the sick.

I. Gentlemen of the faculty of physic, by the authority of different parliamentary statutes, enjoy an exemption from serving on inquests or juries; from bearing armour; from being constables or church-wardens; and from all burdensome offices, whether leet or parochial. These privileges are founded on reasons highly honourable to medical men; and should operate as incentives to that diligent and assiduous discharge of professional duty which the legislature has generously presumed to occupy the time and to employ the talents of physicians and surgeons, in some of the most important interests of their fellow citizens (pp. 61–62).

Thus, in the first articles of Chapters One and Four, Percival links professional prerogatives and duties of professional office. The connection Percival envisions between prerogative and duties emerges more clearly when he discusses the conditions under which a practitioner becomes unfit to enjoy the prerogatives of office.

... As age advances, therefore, a physician should, from time to time, scrutinize impartially the state of his faculties; that he may determine, bona fide, the precise degree in which he is qualified to exercise the active and multifarious offices of his profession. And whenever he becomes conscious that [his abilities have declined]; ... he should at once resolve, though others perceive not the changes which have taken place, to sacrifice every consideration of fame or fortune, and to retire from engagements of business. To the surgeon under similar circumstances, this rule is still more necessary .... Let both the physician and surgeon never forget, that their professions are public trusts, properly rendered lucrative whilst they fulfil them; but which they are bound, by honor and probity, to relinquish, as soon as they find themselves unequal to their adequate and faithful execution (p. 52).

Here Percival makes explicit the Gisbornean presuppositions underlying the first articles of Chapters One and Four, i.e., that there is an tacit understanding between society/government and the profession. Society confers privileges upon the medical professionals (including the right to practice a lucrative occupation) because it believes that so doing furthers "important interests" of its members, specifically their interests in ease, health, and life. Consequently, by accepting their office and its attendant privileges, the medical professional has effectively entered into a contract-like agreement with society, represented by government, to serve his fellow citizens with respect to these interests. The agreement creates a "bond of honor" between practitioner and society, a bond which goes unremarked, except when practitioners are unable to comply with the conditions of the societal-professional contract – as when, for example, senescence impedes the abilities of the practitioner. At that point, practitioners are "bound, by honor and probity to relinquish" the privileges of office, "as soon as they find themselves unequal to their adequate and faithful execution," because to do otherwise is to violate the conditions of the contract between the profession and society.

Percival draws on Gisborne's tacit social contract to develop a conception of the hospital medical practitioner which transcends that of the ordinary employee. Consider his objections to over-crowding and pharmaceutical "economies." In Article XV, Percival asserts that, "it is the duty of the physician or the surgeon to prevent [close wards], as far as lies in his power, by a strict and persevering attention to the whole medical polity of the hospital." In Article VIII, he asserts that "physicians and surgeons should not suffer themselves to be restrained, by parsimonious considerations, from prescribing ... drugs even of a high price, when required in diseases of extraordinary malignity and danger ... no economy, of a fatal tendency, ought to be admitted to institutions" (p. 13). What grounds this "duty" to challenge their employers? There is nothing in the employer – employee relationship per se which justifies it. Nor does the charitable purpose of the hospital supply grounds for such an obligation, for it is not the beneficiary, but the benefactor, who enjoys the right to determine the extent of a benefit; trustees can be beneficent without being munificent.

Yet Percival insists that hospital practitioners have a duty to object if, for reasons of economy, the sick-poor are given ineffective medications or are crowded into poorly ventilated wards. He grounds this obligation neither in the practitioner-patient relationship (the language of rights for hospital patients had yet to be invented) nor in practitioner-patron relationship (since the parties being challenged, the trustees, were the patrons). To justify the obligation of the hospital practitioner to challenge his patron, Percival evokes Gisborne's conception of the medical practitioner as enjoying the prerogatives of station through a tacit social contract. This contractual obligation both creates the important offices of hospital physician and surgeon, while at the same time requiring those who enjoy these offices to attend to the ease, health, and lives of their charges with skill, attention, and fidelity, and – where circumstances require – to challenge trustees on such issues as "close wards" or the stingy provision of medications. To Percival, patrons, whether they be hospital trustees or private parties, are merely a funding mechanism to support medical practitioners in playing a socially defined role. The ultimate obligation of the practitioner, therefore, is neither to the patron nor the patient, it is to society.

Percival's social contractarian, quasi-public servant conception of the practitioner's role explains the otherwise puzzling assortment of "duties" he assigns to them – including his prohibition of surgeons attending at duels (Chapter Four, Articles XII–XIV). For if practitioners are quasi-public servants it is clear that they can not properly play a formal role in an activity officially banned by the government. Again, the quasi-public servant status of the practitioner makes sense of Percival's position on confidentiality – i.e., that a practitioner may not "conceal any part of what he knows" (p. 111) when testifying before a court of law. Quasi-public officials would naturally be expected to place governmental/public interests ahead of those of their patients.

Percival's contractarianism also explains the otherwise puzzling obligation Percival imposes on the private practitioner, in Chapter II, Article XX, of furnishing certificates of illness *free of charge* to anyone seeking an excuse from *civic* obligation, although not from private obligations. The American codifiers, who saw medical practice as an entirely private enterprise, found this duty as incomprehensible. They could not fathom why private practitioners should be expected to provide certificates, without charge, in the case of public duties – especially since Percival leaves them free to charge for certificates of absence from employment and other private activities. Yet, if we presume a Gisbornean societal-professional contract that converts medical practitioners into quasi-public servants, Percival's line of reasoning here seems obvious.

It is perhaps worth remarking that a Gisbornean reading of *Medical Ethics* explains Percival's explanation of his original title for the work: "Medical Jurisprudence." Percival was an Augustan intellectual, who saw imperial Rome as a model for imperial Britain. He not only read the Romans, he was inclined to apply classical Roman conceptions to contemporary endeavors, as is evident both from the content of *A Father's Instructions* and from the way he employs Latin names and precepts in the work. Certain classical natural law theorists, including Justinian, used the term 'jurisprudence' to characterize the process of articulating the moral duties implicit in, or underlying, legal statutes. It is to this tradition which Percival appealed when defending his original title, in his preface to *Medical Ethics*:

> This work was originally entitled "MEDICAL JURISPRUDENCE"; but some friends objected to the term JURISPRUDENCE .... According to the definition of Justinian, however, Jurisprudence may be understood to include moral injunctions as well as positive ordinances. *Juris praecepta sunt haec; honeste vivere; alterum non Laedere; suum cuiquetribuere. Inst Justin*: Lib. I. p. 3." (p. 7).

Freely translated – "These are the precepts of law: to live honorably; not to injure another; to give to each his due." But the puzzle remains, what is the relevance of *law* to medical ethics? He never explains the connection directly, but in his dedication to *Medical Ethics*, he characterizes professional ethics as a synthesis of "knowledge of human nature" and "extensive moral duties." Percival never explicitly states the source of these moral duties. If he is following Gisborne's precedent, however, and deriving moral duties of professionals from a contract which medical practitioners tacitly accept, his original title makes sense. For Percival was literally offering a "jurisprudence" in the Justinian sense: i.e., he was articulating the moral duties implicit in the parliamentary ordinances dealing with medical practice and licensure.

In addition to Gregory and Gisborne, Percival credits a third source which he drew on in writing *Medical Ethics*: the recommendations of a committee of experienced practitioners headed by Dr. Ferriar and Mr. Simmons (a physician and a surgeon at the Manchester Infirmary). The committee had studied regulations and practices already in place at various charitable hospitals. To Percival, it was natural to develop a code by appealing to actual experience. He had a life-long commitment to empiricism which informed both his scientific research and his work on the Manchester census and the Board of Public Health. He disdained theoretical discussions as an irrelevant source of practitioner discord: "For there may be much diversity of opinion, concerning speculative points, with perfect agreement in those modes of practice, which are founded not on hypothesis, but on experience and observation" (Chapter Two, see also Note IX, p. 169).

Percival's faith in "experience and observation" holds the key to his response to the Trustees of the Manchester Infirmary. They had charged him to write a code of ethics for their hospital which would pacify disputes, presumably by developing a basis for collaboration between practitioners with different disciplinary backgrounds. To succeed, such a code had to break decisively with the tradition of tying accountability for patient care to the personal honor of the care-giver – the primary enforcement mechanism in classical virtue ethics, in the ethic of the Hippocratic Oath, and, more importantly, in Gregory.

As has often been remarked, these ethics motivate the physician to provide the patient with good care because failure to do so threatens a loss of standing with himself and others. This ingenious quality assurance mechanism, however, is only effective for solo non-competitive practitioners. For the identification of the character of the practitioner with the quality of care is symmetrical in such ethics – which means that criticisms of the quality of patient care will also be understood as criticisms of the character or ability of the patient's care giver. Or, to put the point another way, insofar as personal honor of the care giver is the motivating force which supports a high standard of patient care, it is also the motivating force which transforms criticisms of patient care into disputes about the personal honor of the care giver.

In the eighteenth century, two developments placed medical practitioners in a position to criticize each others' practices, thereby activating the by and large dormant fractious potential inherent in the identification of personal honor with patient care. The first was the charity hospital, which required physicians, surgeons, and apothecaries to collaborate in providing patient care. Differences between the training and traditions of the three medical specializations guaranteed divergent judgments about the appropriateness of medical treatments. Given the identification of patient care with the practitioner's personal honor, these differences automatically translated into disputes involving personal honor of practitioners. So, in the context of the eighteenth-century hospital, any formula tying the quality of patient care to the practitioner's personal honor became, in effect, a formulary for intractable discord.

The second development engendering practitioner comments on the practices of others – fomenting intra-practitioner disputes – was the relative availability of trained medical practitioners in urban areas. Private patients now had a choice of practitioners. If they became dissatisfied with their progress under one, they could consult another. If, as often happened in these circumstances, a second practitioner changed a medical regimen without consulting the first, e.g., by discontinuing one medication and prescribing another, the ethos which identified honor with this practice left the original practitioner with little choice but to regard the change as a personal insult, a "slander."

Indeed, this was precisely the scenario for most of the better-known medical disputes of the period, including the dispute in which Robert Darwin (son of Erasmus, father of Charles) accused a second practitioner, William Withering, of "slandering" him by discontinuing the medications he had prescribed for a patient, without consultation ([11], pp. 237–239). We know this because the Darwin-Withering correspondence was circulated publicly as a pamphlet. Other disputes went public in the law courts (e.g., the Fothergill-Leeds dispute [11], pp. 241–243), while still others were published by the local press – e.g., the disputes discussed in Chapters One and Two, and, to cite another example, the Simmons-Hall/White dispute that racked the Manchester Infirmary in the 1790s ([18], pp. 237–241).

It was this later dispute that prompted the trustees of the Infirmary to turn to Percival and which his code addressed directly. Although Percival himself was a virtue theorist, and although in his dedication he refers to virtue theory, the dispute problem was essentially unresolvable within the compasses of a virtue ethic. Percival could resolve the dispute problem only by turning away from personal honor as a mechanism of enforcement and looking elsewhere. As a good empiricist, he turned to the regulations collected by Dr. Ferriar and Mr. Simmons, which characterize a collaborative decision-making and dispute resolution process that would today be called "rounds."

On the basis of Ferriar-Simmons' recommendations, Percival stipulates that before any treatment plan is developed for a hospitalized patient, and before any major change in a treatment plan is decided, a formal conference (i.e., "rounds") should be held which is open to any practitioners with an interest in a case. At this conference, practitioners should state their views in *inverse* order of seniority (Chapter One, Article

XIX) – temporarily suspending hierarchy to ensure a frank and open discussion and mitigating the influence of superordinate's opinions on his/her subordinate. The treatment plan to be adapted is that which is recommended by the *majority* of practitioners, even if it is not embraced by the senior physician or surgeon – as Percival puts it "a majority should be decisive," (p. 20). Percival allows two pragmatic exceptions to majority rule: "If the numbers be equal, the decision should rest with the physician or surgeon under whose care the patient is placed" (note again, not the senior physician/surgeon); he also allows this attending physician/surgeon the right of "making such variations in the mode of treatment, as further contingencies may require, or as further insight into the nature of the disorder may show to be expedient" (Article XIX, p. 20).

Percival's faith in the superiority of collective over individual decisionmaking permeates *Medical Ethics*. At every point, he attempts to minimize the importance of the hospital hierarchy and to subordinate individual decisionmaking to collective decisionmaking and/or scrutiny. In Chapter One, Article XV, for example, Percival recommends developing a hospital register (i.e., a system of medical record keeping) to determine "the comparative success of their hospital and private practice." Article XVII, however, stipulates that such investigations are best conducted by a "committee of the faculty" because "the exertions of individuals, however benevolent or judicious, are opposed by those who have not been consulted; and prove inefficient by wanting the collective energy of numbers."

In Chapter One, Article XII, in a discussion of what would today be called "clinical experimentation on human subjects," Percival stipulates that: "no such trials shall be instituted, without a previous consultation of the physicians or surgeons with the "gentlemen of the faculty." In other words, individual investigators are to subject their reasons for the proposed experiment to the collective assessment of the faculty, who are to examine "scrupulously and conscientiously," whether the experiment is based on "sound reason and just analogy, or well authenticated facts."

A comparison with Gisborne highlights the radicalness of Percival's subordination of *individual* to *collective judgment*. Gisborne is, if anything, even more concerned than Percival about the possibility of "unprincipled" practitioners, performing reckless experiments on hospital patients, reasoning that "in the populous wards of an infirmary the ill success of an adventurous trial is lost in the crowd of unfortunate and fortunate events; and even if it should terminate in the death of an obscure, indigent, and quickly forgotten individual, little if any disadvantage results to the credit and interest of the physician among his wealthy employers" ([7], p. 407). Yet, so deeply entrenched was the tradition of individualistic morality that it never occurred to Gisborne to curb "rash, hastily adopted, ... ignorant ... careless ... obstinate experimenters" by having them submit their projects to collective review by the faculty of the institution. Gisborne relies entirely on the experimenter's conscience; Percival takes the more radical step of subjecting the individual's conscience to peer review.

In Article X, Chapter One, which bears the title "Professional charges to be made only before a meeting of the faculty" (p. xi), Percival speaks directly to the problem of disputes:

> No professional charge should be made by a physician or a surgeon, either publicly or privately, against any associate, without previously laying the complaint before the gentlemen of the faculty belonging to the institution, that they may judge concerning the reasonableness of the grounds, and the measures to be adopted (p. 14).

In Chapter Two, Percival extends this obligation into private practice (explicitly employing the Gisbornean argument of a the "tacit compact"):

> XIII. The Esprit de Corps is a principle of action founded in human nature, and when duly regulated, is both rational and laudable. Every man who enters into a fraternity engages, by tacit compact, not only to submit to the laws, but to promote the honor and the interests of the association, so far as they are consistent with morality, and the general good of mankind. A physician, therefore, should cautiously guard against whatever may injure the general respectability of his profession; and should avoid all contumelious representations of the faculty at large; all general charges against their selfishness or improbity; and the indulgence of an affected or jocular skepticism, concerning the efficacy and utility of the healing art (pp. 45–46).

Percival also holds that the tacit intra-practitioner compact also obligates private practitioners to submit their disputes to arbitration without seeking public vindication of their personal honor.

> XXIV ... [When] controversy, and ... contention ... occur, and can not be immediately terminated, they should be referred to the arbitration of a sufficient number of physicians or surgeons, ... or to the orders of both collectively .... But neither the subject, nor the adjudication, should be communicated to then public; as they may be personally injurious to the individuals concerned, and can hardly fail to hurt the general credit of the faculty (p. 46).

In Chapter Two, Article IV Percival requires that before "whistleblowers" provoke a dispute with other practitioners, they too should subject their individual judgments to intra-subjective scrutiny.

Officious interference, in a case under the charge of another, should be carefully avoided .... Yet though the character of a professional busy-body, whether from thoughtlessness or craft, is highly reprehensible, there are occasions which not only justify but require a spirited interposition. When artful ignorance grossly imposes upon credulity; when neglect puts to hazard an important life; or rashness threatens it with still more danger; a medical neighbor, friend, or relative, apprized of the facts will justly regard his interference as a duty. But he ought to be careful that the information, on which he acts, is well founded; that his motives are pure and honourable; and that his judgment of the measures pursued is built on experience and practical knowledge, not on speculation or theoretical differences of opinion .... In general ... a personal and confidential application to the gentlemen of the faculty concerned should be the first step taken and afterwards, if necessary, the transaction may be communicated to the patient or to his family (pp. 32–33).

Thus, even though this is a private practice and not a hospital matter, it is impermissible for the practitioner to discuss this matter directly with the patient or family concerned without *first* presenting the case to the faculty of the hospital, whose disinterested and intra-subjective judgment will presumably act as a constraint on subjective bias.

Notice that despite Percival's predilection for collective over individual judgment, he eventually allows the individual practitioner his independence, even at the risk of engendering public disputes, *viz.*, "if necessary, the transaction may be communicated to the patient or his family" [33]. Thus the sole case in which Percival refuses to constrain individual practitioner autonomy is when it is asserted in the interests of the patient.

Percival's solution of the dispute problem, therefore, involves the suspension of hierarchy, the substitution of consultation for individual judgments, and the use of the hospital faculty as a board which sets medical policy and provides a venue for the arbitration and adjudication of intra-practitioner disputes. Yet these innovations alone could not resolve the problem of medical disputes unless they are coupled with a more radical departure: divorcing accountability from personal honor. In Chapter Two, Percival offers an entirely novel conception of practitioner accountability, albeit one that the medical sociologist, Charles Bosk has shown is standard practice in contemporary teaching hospitals.

XXVIII. At the close of every interesting and important case, especially when it hath terminated fatally, a physician should trace back, in calm reflection, all the steps which he had taken in the treatment of it. This review of the origin, progress, and conclusion of the malady; of the whole curative plan pursued, and of the particular operation of the several remedies employed, as well as of the doses and periods of time in which they were administered, will furnish the most authentic documents, on which individual experience can be formed. But it is in a moral view that the practice is here recommended; and it should be performed with the utmost scrupulous impartiality. Let no self-deception be permitted the retrospect; and if errors either of omission or commission, are discovered, it behoves that they should be brought fairly and fully to mental view. Regrets may follow, but criminality will thus be obviated. For the good intentions, and the imperfection of human skill, which cannot anticipate the knowledge that events alone disclose, will sufficiently justify what is past, provided this failure be made conscientiously subservient to future wisdom and rectitude in professional conduct (pp. 48–49).

Most conceptions of accountability turn on two crucial variables: the *intent* of the agent, and the *consequences* of the agent's actions. Normally, when both intent and consequences are bad (i.e., if harm is both intended and inflicted) the agent is considered criminal, in the sense of being both blameworthy and punishable. On the other hand, where a harmful act is unintended, only consequences count, and the agent is considered non-criminal but nonetheless blameworthy, at least to the extent that the agent is held liable (for malpractice) and so must compensate the victim for the damages suffered. In hospital medicine, as Percival points out, the practitioner's "good intentions" can typically be taken for granted, yet despite these good intentions, patients often have bad, even fatal outcomes, *even though the practitioner is not guilty of any malpractice*. The reason is that the "imperfections of human skill" are such that no practitioner can guarantee good outcomes. Bad outcomes, therefore, cannot be the standard by which medical practice is to be judged.

What is the standard of accountability to which medical practitioners are to be held? Percival argues that bad outcomes can be forgiven, if the practitioner reviews the entire case, impartially, with "no self-deception," and frankly admits "errors either of omission or commission." He makes one important *proviso*: "provided the failure be made conscientiously subservient to future wisdom and rectitude in professional conduct," i.e, provided that the mistake is not repeated in the future.

What Percival is recommending here is a form of "forgive and remember" accountability, to use Bosk's felicitous turn of phrase, that holds the practitioner morally accountable, not for outcomes consequent to past actions, but for failure to remember and learn from past mistakes. What is unforgivable is not the death of a patient, even if the death was avoidable, it is the failure to learn from this death and make it "subservient to future wisdom and rectitude." The unprofessional practitioner is not one who practices poorly on occasion, but one who refuses to admit such errors and to learn from them. Thus, on Percival's redefinition of professional virtue, consultation is transformed into a helpful corrective, a forum in which the practitioner displays professional virtue by admitting error rather than quarrelling with critics and crying "slander." For the only accusation which could "slander" practitioners' reputation would be that they had failed to frankly admit errors or to learn from past mistakes.

Read in the context of the eighteenth century, Percival's code is no cipher. He forthrightly acknowledges the sources he drew on in writing the code – Gregory, Gisborne, Ferriar and Simmons – and explains the problems he was addressing. Like many moral theorists working at the end of the eighteenth century (e.g., Jeremy Bentham's *Introduction to the Principles of Morals and Legislation* (1789) and Immanuel Kant's *Groundwork of the Metaphysics of Morals* (1785)), in attempting to deal with these problems he was forced to transcend the first-person ethics of personal honor and virtue which had dominated secular ethics for centuries. So Percival, working quite independently, spent the last decade of the eighteenth century wrestling with a similar problem with a narrower focus – the problem of devising an intersubjective code of professional conduct appropriate to both hospital and private medical practice. As an Enlightenment reformer, he also strove for equitable treatment of hospital patients and this, in turn, required him to maintain the independence of the practitioner from hospital patrons.

Nothing in the existing literature, not Gregory, not Gisborne, and certainly not Hippocrates, could resolve all of these problems. So Percival was forced to break from first-person approaches to medical morality and to ground professional duty in a *societal-professional contract which permitted and encouraged collaborative practice*, regulated by the profession itself through its representatives, the faculty/staff of a teaching hospital. Percival's code writes the epitaph for individualistic virtue ethics in medicine, supplanting them with the ideal of a collaborative profession committed to the development of a scientific, empirically-based medicine, dedicated to treating the sick, whether rich or poor, and subjecting the treatment decisions of independent practitioners to intra-subjective validation by their peers. Percival's prescient vision of the nature of a profession captured the imagination of the American codifiers, who embedded it in their various codes, creating the basis of the Anglo-American medical profession and its morality.

ACKNOWLEDGEMENTS

I should like to thank Dorothy and Roy Porter, Larry McCullough, and Julie Walter for their many helpful comments on various drafts of this paper. I am especially indebted to Marianne Snowden for her advice and assistance in the preparation of the manuscript.

Union College

NOTES

[1] Editions of Percival's Code(s) of *Ethics*

Medical Jurisprudence	Manchester, 1794
Medical Ethics	J. Johnson, Manchester, 1803
Anonymous, ed.	J. Johnson, London, 1827
Greenhill, ed.	J.H. Parker, Oxford, 1849
....	
Chauncey Leake, ed.	William & Wilkins, Baltimore, 1927
Chester Burns, ed.	Robert E. Kreiger, Huntington N.Y., 1975 (reprint 1927, Leake edition)
Edmund Pellegrino, ed.	Classics of Medicine Library, Birmingham, Alabama, 1985 (facsimile of 1803 edition)
Extracts from the Medical Ethics of Thomas Percival	Lexington, Kentucky, 1821 Philadelphia, Pennsylvania, 1823

(The dotted line indicates the transition from British to American editions of *Medical Ethics.* No British edition of Percival is published after 1849, but twentieth-century American publishers reissue *Medical Ethics*, treating it as a significant historical document and publishing a facsimile edition.) All references in this work are to the 1803 edition, of *Medical Ethics*, unless otherwise indicated.

[2] In his detailed study of the codes of medical morality in the United States, Chester Burns accords the honor of being *first* to the March 1808 Boston Medical Association code.

Known as the Boston Medical Police, this code had been prepared by a committee of doctors who claimed that they used the writings of Gregory, Percival, and Rush. Actually, all of the precepts in the Boston Medical Police could be found in the second chapter of Percival's *Medical Ethics*, the chapter that discussed such situations as consultations, arbitration of differences, interferences with another's practice, fees, and seniority amongst practioners .... [T]he Boston Medical Police became the model of codes of medical ethics adopted between 1817 and 1842 by at least thirteen societies in eleven states, New York not included ([5] p. 302).

The New York State Medical Society and the Medico-Surgical society of Baltimore, drafted codes in the 1830s, although independent of the Boston Medical Police these codes, too, drew most of their inspiration, and much of their language, from Percival. A similar process went on in Canada, where the newly formed Canadian Medical Association (1867) formulated a Code of Ethics, strongly influenced by the American model ([6] p. 966).

[3] Among the other sources were writings by John Gregory (1724–1773) [9] and his students, including his son, James Gregory (1753–1821), and two Americans, Samuel Bard (1742–1821) and Benjamin Rush (1745–1813).

[4] In the first half of the nineteenth century, American practitioners were in the process of organizing themselves. In doing so they eschewed the British pattern of organizing medicine according to colleges mirroring a tripartite medical division of labor (into physicians, surgeons and apothecaries); the division had little currency in America, so they opted instead to organize medical societies by geographic regions (cities, states). Each of these nascent medical organizations required a charter, and it became standard to incorporate into these charters a section dealing with medical ethics, or etiquette, or police (a Scottish term, used by Gregory). As Leake notes, Percival's pre-codified *Medical Ethics* was ideal for these purposes. Here, ready-to-hand, was an Enlightenment code of medical ethics, drafted by an English philosopher, which the Americans, who saw themselves as heirs to the Enlightenment, could use to state their ideals for themselves. Thus the singular use to which Percival's *Medical Ethics* lent itself was as a source-book for those drafting codes of ethics for city and state medical societies.

REFERENCES

1. American Medical Association: 1989, *Current Opinion and Principles of Medical Ethics*. American Medical Association, Chicago.
2. Berlant, J.: 1975, *Profession and Monopoly: A Study of Medicine in the United States and Great Britain*, University of California Press, Berkeley.
3. Bosk, C.: 1979, *Forgive and Remember*, University of Chicago Press, Chicago.
4. Burns, C.: 1977, "Reciprocity in the Development of Anglo-American Medical Ethics, 1765–1865", in C. Burns (ed.), *Legacies in Ethics and Medicine*, Science History Publications, New York, pp. 300–307.
5. Edelstein, L.: 1967, "The Professional Ethics of the Greek Physician", in Owsei and C. Lilian Temkin (eds.), *Ancient Medicine: Selected Papers of Ludwig Edelstein*, Johns Hopkins University Press, Baltimore, pp. 319–348.
6. Gisborne, T.: 1795, *Principles of Moral Philosophy Investigated and Briefly Applied to the Constitution of Civil Society*, B and J White, London.
7. Gisborne, T.: 1794, *An Enquiry into the Duties of Men in the Higher and Middle Classes of Society in Great Britain Resulting from their Respective Stations, Professions and Employment*, B. and J. White, London.
8. Gregory, J.: 1817, *Lectures on the Duties and Qualifications of a Physician*, 2nd

ed., M. Carey & Son, Philadelphia.

9. Hume, D.: 1777, *Enquires Concerning Human Understanding and Concerning the Principles of Morals*, 2nd. ed., L. Selby Bigge (ed.), Oxford University Press, London (1902).
10. Hutcheson, F.: *Inquiry into the Original of Our Ideas of Beauty and Virtue*, London,
11. King, L.: 1958, *The Medical World of the Eighteenth Century*, University of Chicago Press, Chicago.
12. Larkey, S.: 1952, "The Hippocratic Oath in the Elizabethan Period", *Bulletin of the History of Medicine*. 26, 1–31.
13. Leake, C., ed.: 1927, *Percival's Medical Ethics*, Williams and Wilkins, Baltimore.
14. Pellegrino, E.: 1984, "The Virtuous Physician and Ethics of Medicine", in E. Shelp (ed.), *Virtue and Medicine*, D. Reidel, Dordrecht.
15. Percival, E.: 1807, *Memoirs of the Life and Writings of Thomas Percival*, J. Johnson, London.
16. Percival, T.:1985, *Medical Ethics*, in Pellegrino, E. (ed.), *The Classics of Medicine Library*, Gryphon Editions Ltd., Birmingham.
17. Percival, T.: 1807, *The Works, Literary, Moral and Medical of Thomas Percival*, J. Johnson, London.
18. Pickstone, J.V. and Butler, S.V.: 1984, "The Politics of Medicine in Manchester, 1788–1792", *Medical History* 28, 227–249.
19. Reich, W. (ed.): 1978, "Medical Ethics, History of: North America; Seventeenth to Nineteenth Century", *Encyclopedia of Bioethics*, Free Press, New York, p. 966.
20. Rush, B.: 1794, "Duties of a Physician", *Medical Inquiries and Observations, Volume I*, Philadelphia, appendix.
21. Selby-bigge, L.A.: 1964, *British Moralists: Being Selections from the Writers Principally of the Eighteenth Century*, Bobbs-Merrill, New York.
22. Shryock, R.: 1936, *The Development of Modern Medicine*, University of Wisconsin Press, Madison.
23. Smith, A.: 1976, D. D. Raphael, A. L. MacFie (eds), *The Theory of Moral Sentiments*, The Clarendon Press, Oxford.
24. Veatch, R.: 1981, *A Theory of Medical Ethics*, Basic Books, New York.
25. Waddington, I.: 1975, "The Development of Medical Ethics – A Sociological Analysis", *Medical History* 19, 36–51.
26. Welborn, M.: 1977, "The Long Tradition: A Study in Fourteenth Century Medical Deonotology", in C. Burns (ed.), *Legacies in Ethics and Medicine*, Science History Publications, New York, pp. 204–217.

NOTES ON CONTRIBUTORS

Robert Baker, Department of Philosophy, Union College, Schenectady.

Tom L. Beauchamp, Department of Philsophy, and Kennedy Institute of Ethics, Georgetown University, Washington, D.C.

Mary E. Fissell, Institute of the History of Medicine, The Johns Hopkins University, Baltimore.

Johanna Geyer-Kordesch, Wellcome Unit for the History of Medicine, University of Glasgow, Glasgow.

David Harley, independent scholar and historian.

Laurence B. McCullough, Center for Ethics, Medicine, and Public Issues, Baylor College of Medicine, Houston.

John Pickstone, Wellcome Unit for the History of Medicine, The University, Manchester.

Dorothy Porter, Department of History, Birkbeck College, University of London, London.

Roy Porter, Wellcome Institute for the History of Medicine, London.

R. Baker, Dorothy Porter and Roy Porter (eds), The Codification of Medical Morality, 213.

INDEX

R. Baker, Dorothy Porter and Roy Porter (eds), The Codification of Medical Morality, 215–230.

23. E.E. Shelp (ed.): *Sexuality and Medicine.*
Vol. II: Ethical Viewpoints in Transition. 1987
ISBN 1-55608-013-1; Pb 1-55608-016-6
24. R.C. McMillan, H. Tristram Engelhardt, Jr., and S.F. Spicker (eds.): *Euthanasia and the Newborn.* Conflicts Regarding Saving Lives. 1987
ISBN 90-277-2299-4; Pb 1-55608-039-5
25. S.F. Spicker, S.R. Ingman and I.R. Lawson (eds.): *Ethical Dimensions of Geriatric Care.* Value Conflicts for the 21th Century. 1987
ISBN 1-55608-027-1
26. L. Nordenfelt: *On the Nature of Health.* An Action- Theoretic Approach. 1987
ISBN 1-55608-032-8
27. S.F. Spicker, W.B. Bondeson and H. Tristram Engelhardt, Jr. (eds.): *The Contraceptive Ethos.* Reproductive Rights and Responsibilities. 1987
ISBN 1-55608-035-2
28. S.F. Spicker, I. Alon, A. de Vries and H. Tristram Engelhardt, Jr. (eds.): *The Use of Human Beings in Research.* With Special Reference to Clinical Trials. 1988 ISBN 1-55608-043-3
29. N.M.P. King, L.R. Churchill and A.W. Cross (eds.): *The Physician as Captain of the Ship.* A Critical Reappraisal. 1988 ISBN 1-55608-044-1
30. H.-M. Sass and R.U. Massey (eds.): *Health Care Systems.* Moral Conflicts in European and American Public Policy. 1988 ISBN 1-55608-045-X
31. R.M. Zaner (ed.): *Death: Beyond Whole-Brain Criteria.* 1988
ISBN 1-55608-053-0
32. B.A. Brody (ed.): *Moral Theory and Moral Judgments in Medical Ethics.* 1988
ISBN 1-55608-060-3
33. L.M. Kopelman and J.C. Moskop (eds.): *Children and Health Care.* Moral and Social Issues. 1989 ISBN 1-55608-078-6
34. E.D. Pellegrino, J.P. Langan and J. Collins Harvey (eds.): *Catholic Perspectives on Medical Morals.* Foundational Issues. 1989 ISBN 1-55608-083-2
35. B.A. Brody (ed.): *Suicide and Euthanasia.* Historical and Contemporary Themes. 1989 ISBN 0-7923-0106-4
36. H.A.M.J. ten Have, G.K. Kimsma and S.F. Spicker (eds.): *The Growth of Medical Knowledge.* 1990 ISBN 0-7923-0736-4
37. I. Löwy (ed.): *The Polish School of Philosophy of Medicine.* From Tytus Chałubiński (1820–1889) to Ludwik Fleck (1896–1961). 1990
ISBN 0-7923-0958-8
38. T.J. Bole III and W.B. Bondeson: *Rights to Health Care.* 1991
ISBN 0-7923-1137-X
39. M.A.G. Cutter and E.E. Shelp (eds.): *Competency.* A Study of Informal Competency Determinations in Primary Care. 1991 ISBN 0-7923-1304-6
40. J.L. Peset and D. Gracia (eds.): *The Ethics of Diagnosis.* 1992
ISBN 0-7923-1544-8
41. K.W. Wildes, S.J., F. Abel, S.J. and J.C. Harvey (eds.): *Birth, Suffering, and Death.* Catholic Perspectives at the Edges of Life. 1992 ISBN 0-7923-1547-2

42. S.K. Toombs: *The Meaning of Illness.* A Phenomenological Account of the Different Perspectives of Physician and Patient. 1992 ISBN 0-7923-1570-7
43. D. Leder (ed.): *The Body in Medical Thought and Practice.* 1992 ISBN 0-7923-1657-6
44. C. Delkeskamp-Hayes and M.A.G. Cutter (eds.): *Science, Technology, and the Art of Medicine.* European-American Dialogues. 1993 ISBN 0-7923-1869-2
45. R. Baker, D. Porter and R. Porter (eds.): *The Codification of Medical Morality.* Historical and Philosophical Studies of the Formalization of Western Medical Morality in the Eighteenth and Nineteenth Centuries, Volume One: Medical Ethics and Etiquette in the Eighteenth Century. 1993 ISBN 0-7923-1921-4

KLUWER ACADEMIC PUBLISHERS – DORDRECHT / BOSTON / LONDON

MIX
Papier aus verantwortungsvollen Quellen
Paper from responsible sources
FSC® C105338

If you have any concerns about our products,
you can contact us on
ProductSafety@springernature.com

In case Publisher is established outside the EU,
the EU authorized representative is:
Springer Nature Customer Service Center GmbH
Europaplatz 3, 69115 Heidelberg, Germany

Printed by Libri Plureos GmbH
in Hamburg, Germany